Urban and Environmental Economics

T0203755

The importance of urban spaces to environmental protection and social issues is well established, particularly within the working lives of planners, developers, architects, building designers and quantity surveyors. Those new to this industry, or preparing to join it, must have an understanding of how their environmental and social responsibilities relate to their professional responsibilities in economic terms.

Designed as an introductory textbook, *Urban and Environmental Economics* provides the background information to understand crucial economic concepts and techniques. A broad range of theories of the natural and built environment and economics are explained, helping the reader to develop a real understanding of the topics that influence this subject, such as:

- the history of economic thought and perspectives of urban and environmental economics;
- market forces, market failure and externalities in the built and natural environment;
- introduction to cost-benefit analysis;
- macroeconomic tools and policy;
- issues in urban space: city models and land use;
- environmental resources and use in urban space: pollution control, natural resource and energy economics, resource valuation;
- international contemporary challenges: climate change, globalisation, sustainable urban development, appraisal, evaluation and assessment, and informatics.

Illustrated throughout and with lists of key concepts in every chapter, this book is ideal for students at all levels who need to get to grips with the economics of the environment within an urban and built environment context. The book is particularly useful to those studying planning, architecture, land economy, environmental management, property, geography, real estate, construction, housing, regeneration and public policy.

Graham Squires is a Senior Lecturer at the University of the West of England (UWE), Bristol. His research interests include economics (urban and environmental), development, real estate, spatial planning, housing, neighbourhoods and regeneration. He has written academic books and journal articles, in addition to working on various consultancy projects in policy and practice. His previous book, *Introduction to Building Procurement*, was published by Routledge in 2011.

Urban and Environmental Economics

An introduction

Graham Squires

Routledge
Taylor & Francis Group

LONDON AND NEW YORK

First published 2013
by Routledge
2 Park Square, Milton Park, Abingdon, Oxon, OX14 4RN

Simultaneously published in the USA and Canada
by Routledge
711 Third Avenue, New York, NY 10017

Routledge is an imprint of the Taylor & Francis Group, an informa business

© 2013 Graham Squires

The right of Graham Squires to be identified as author of this work has
been asserted by him in accordance with sections 77 and 78 of the
Copyright, Designs and Patents Act 1988.

British Library Cataloguing in Publication Data
A catalogue record for this book is available from the British Library

Library of Congress Cataloging-in-Publication Data
Squires, Graham.
Urban and environmental economics : an introduction / Graham Squires.
p. cm.
Includes bibliographical references and index.
1. Urban economics. 2. Environmental economics. I. Title.
HT321.S69 2013
330.9173'2–dc23
2012009071

ISBN: 978-0-415-61990-5 (hbk)
ISBN: 978-0-415-61991-2 (pbk)
ISBN: 978-0-203-82599-0 (ebk)

Typeset in Bembo by
FiSH Books Ltd, Enfield

Contents

List of illustrations

Figures

Tables

Photographs

All photographs © the author

Acknowledgements

Thanks are given to all family, friends, students and colleagues who have helped shape the writing of this book. Special thanks are given to Fiona for her loving support and patience.

For Tony Squires (1943–2002).

Love and peace.

Chapter 1

What is urban and environmental economics?

Vancouver, Canada

a. Overview

This text on urban and environmental economics introduces economics to the interlocking paradigms of both urban and environmental issues. Disciplines of both urban economics and environmental economics have often tended to take a separate and insular view from each other – it is one intention of this book to unify such thinking as well as extending thought into this field. Whilst doing this, the text more pragmatically demonstrates what urban and environmental economics entails within theory, concepts and practice. Furthermore, an introduction of techniques and tools within the subject will be outlined for the reader to use within both research and further studies.

Conceptually, the focus of the book involves three interlocking strands that are viewed through an economic lens: the built environment; urban issues; and environmental resources (Figure 1.1). The built environment strand will connect to the spatial context of study; moreover, built-up areas at varying spatial scales will form the canvas on which discussion is expressed. This, for instance, involves analysis of neighbourhood effects within a city, or develops analysis of the agglomeration of cities within a larger urban conurbation or metropolis. Within these geographies, the economic concepts to be unearthed are those that integrate a multitude of themes within the urban issue strand. Themes of issues attached to urban geographies include education, such as examining how educational attainment is distributed over space; or those such as housing, where analysis involves an exploration of how the value of a neighbourhood correlates to household income and wealth. The third strand of environmental resources will draw together both urban issues and the spatial context of the built environment. For instance, the implications of de-urbanising and urbanising areas will be a multitude of needs and wants, which will have to be met, if possible using a scarce amount of available environmental resources. An urbanising world city will have wants such as building materials and energy needs, these will be met in part or in entirety, depending upon the availability of such resources – and thus the economic choices and decisions will play a significant role in how urban areas develop.

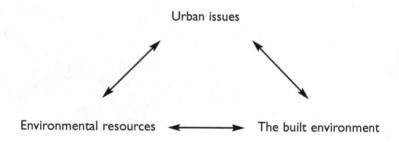

Figure 1.1 Three interlocking strands of urban and environmental economics
Source: Author

Within the three main strands, basic economic concepts for resource allocation are introduced that will be relevant to the planning, valuation and management of shared spaces. Economic concepts to be explored for application on urban and environmental matters include key issues such as considering the limits to growth and how choice is being played out within the built environment given the scarce environmental resources available. With respect to scarcity and choice, economic tools can be applied to provide more technical measurement of urban and environmental issues. Examples of economic tools include the use of cost-benefit analysis (CBA), and hedonic modelling that can provide more empirical evidence to support both arguments and decision-making in this field. Understanding and competence in these tools and models can aid in their use when applying them to both general and specific case studies and examples. For instance, more general comparative CBA could demonstrate the common and differentiating features of introducing a road or rail bypass for an urban area (e.g. London Cross-Rail), whereas the specific introduction of a housing estate to relieve residential pressures in an urban area could make use of a hedonic house price model that measures the dynamics (involving time) and magnitude (steepness of change) of particular independent variables such as education and health in relation to changes in a dependent variable such as house price.

b. Relevance and importance

At the forefront of urban and environmental economics is the pressing issue that urban areas are set to become the focus for the global effort to curb climate change. It is argued that the world's cities are responsible for about 70 per cent of emissions, yet they only represent 2 per cent of the planet's land cover (UN Habitat, 2011). More dramatically, it has been argued that there will be a collision between climate change and urbanisation if no action is taken. According to a recent UN report, an estimated 59 per cent of the world's population will be living in urban areas by 2030. Plus, every year, the number of people who live in cities and towns grows by 67 million – 91 per cent of this figure is being added to urban populations in developing countries (ibid.). Reasons why urban areas are energy-intensive is due to increased transport use, heating and cooling homes and offices and economic activity to generate income.

As well as cities' contribution to climate change, towns and cities around the globe were also vulnerable to the potential consequences. These include an increase in the frequency of warm spells/heatwaves over most land areas, a greater number of heavy downpours, a growing number of areas affected by drought and an increase in the incidence of extremely high sea levels in some parts of the world. Economic problems will be generated from these physical risks posed by future climate change, with, for instance, some urban areas facing difficulties in affording to provide basic services following change. For example, these changes will affect water supply, physical infrastructure, transport, ecosystem goods and services, energy provision and industrial production. At a more microscale, local economies will be disrupted and populations will be stripped of their assets and livelihoods. Disparities of income and wealth at a macroscale can also be extrapolated with arguments that those poorer urban areas are doubly disadvantaged, with a lack of wealth to mitigate against climate change whilst also being subject to its damaging effects.

c. Introducing urban and environmental economics

In understanding what urban and environmental economics entails, an overview of its importance as a subject area and its relevance to both knowledge and practice will need to be determined. Its evolution by integrating economics to both urban issues and environmental concerns within the built environment involve both historical and current economic thinking within the fields of urban studies and environmental science. As such, the attributes and historical context of urban and environmental economics will, initially, be revealed as a subject in itself and in its application to policy.

Isolated constituent parts of what is meant by urban economics and environmental economics are defined and illustrated while highlighting the common ground in which the two disciplines are intertwined. The concepts of space using the urban or built environment are also introduced to provide clarity to 'where' the focus of discussion is being placed. This is particularly important, as not only are processes operating in urban areas, but there is connectivity between built environments situated in space (e.g. roads, rail networks, airports), which have demands on natural resources. In considering these connectors of the built environment, economic aspects can be attached, such as how much and what value of natural resources (e.g. energy) is supplied external to an urban area (e.g. a power station) and being transmitted and consumed internally by a particular densely populated city or region. Natural resources as part of this systematic thinking is also introduced as a perspective in the subject – especially in terms of how environmentalism and the sustainability agenda operate in relation to urban areas, plus further coherence in distinguishing between what is meant by the 'natural' and 'built' environment. For instance, urban green space (i.e. parks) and green belts as the natural environment may be conceived differently, plus changes in land use (e.g. to brownfield) may render them sites that are no longer considered 'natural'.

Different paradigms in what is considered environmental or urban in economics should be clarified and are also considered in this text (Chapter 2). Two significant and distinctive types of thinking are more modern western environmental resource economic (ERE) paradigms and more ecological economic (EE) paradigms. The former approach, environmental resource economics, tends to consider largely classical economic ideas in relation to the environment, such as internalising into the market the external cost of pollution. The latter paradigm, ecological economics, tends more to integrate elements of economics, ecology, thermodynamics, ethics and a range of other natural and social sciences to provide an integrated and biophysical perspective on environment–economy interactions (Van den Bergh, 2000). Links to ecological economics will be drawn on in this text, as an ecological approach will introduce critical linkages to how urban areas produce and consume environmental resources.

The appropriation of economics to both the built and natural environment is no doubt complex and generates different and often vociferous opinion on what approaches 'are' and 'ought' to be carried out by economists. Urban and environmental economics according to economists, both individually and institutionally, is the focus of Chapter 3. This outline of significant thinkers integrates both norma-tive and empirical analytical dimensions. Normative statements are primarily non-falsifiable statements or value judgements of what should or ought to be – a normative statement could argue that rapid urban sprawl should be halted as it is detrimental to sustainable development. Positive statements are more empirically driven to provide information that can be made falsifiable or tested for its basis in 'fact' – an empirical statement could be that the global proportion of urban popu-lation rose dramatically from 13 per cent (220 million) in 1900, to 29 per cent (732 million) in 1950, to 49 per cent (3.2 billion) in 2005 (UN, 2005).

Key writers to introduce urban and environmental economic thinking include Thomas Malthus (1766–1834). The key concern for Malthus was the rapid increase in population to urban areas as a process of urbanisation, and he intonated that while the population grows exponentially, the supply of resources such as food and shelter can only grow linearly. This in turn puts a limit on the size of population a given area can support and, as a result, generates a problem of the growth and distribution of resources within the built environment. As well as Malthus, other classical economists have been attached to urban and environmental thought. An influential writer is David Ricardo (1772–1823) who made a significant connec-tion between the prices of natural resource produce, such as food, with economic concepts, such as comparative advantage and rents.

Infinite economic advance and the ability to feed populations took a different direction with the influential writing of John Stuart Mill (1806–1873). Mill, who initially took a libertarian free-market slant, viewed economic growth in tune with natural cycles, where growth was not an endless process but one that will eventu-ally reach a lasting dynamic equilibrium. Attachment of these cycles of economic growth was at the forefront of thinking for Mill during a period where material advancement held sway. Later writing incorporated social welfare ideas into economic thought, where, for instance, well-being could be maximised for all when the greatest consumption capacity is for the greatest number of individuals. Social welfare in urban economics is particularly important in contemporary thought, with a concern for internal social costs, such as those attached to concen-trated poverty, largely located in urban areas. This in turn brings focus away from pure economic growth models and towards issues such as redistribution of income and wealth within and between urban areas.

Arthur Pigou (1877–1959) is discussed with respect to social concerns and envi-ronmental resources in the urban environment. Pigou's work on income distribution within welfare economics is important, especially as it was argued that a greater balance of wealth distribution could increase greater consumption and growth. His work on externalities also firmly fixed his influence within the realm of environmental economics. For instance, as a corrective measure to counter external costs from pollution, a series of 'Pigovian taxes', such as petrol or carbon taxes (also referred to as ecotaxes or the Green Tax Shift), have been endorsed.

More recent institutional approaches have been developed by groups such as the Club of Rome (1968), which involved former heads of state, UN bureaucrats, high-level politicians and government officials, diplomats, scientists, economists and business leaders from around the globe. Their fundamental impact, with regards to environmental economics, was to publish on the limits to growth as a result of rapidly growing populations and finite resource supplies (Meadows *et al.*, 1972). It can therefore be recognised that some of the larger fundamental economic questions regarding scarce environmental resources and growing population demands keep Malthus as relevant today as in the eighteenth century.

At a more general level, urban and environmental economics are primarily concerned with the basic economic problem of unlimited wants and scarce resources (Chapter 4). For this book, the environmental resources, 'wants', are situated with respect to urban spaces. For instance, meeting the 'want' of water in urban areas will be dependent on how society decides to produce and distribute the resource. These decisions centre on choice (if there is a choice), and if resources are scarce, the production and distribution will be made via a system of choosing alternative uses, through which the need could be met.

Choice in the use of resources for production is important in understanding how society decides the what, how and to whom in an urban environment. If there is a choice it will mean that there is an opportunity forgone on the alternative – this is what is referred to as an 'opportunity cost'. If there is a choice to use public funds for a new school, there will be an opportunity cost of what alternative funds could have been used – for example to build a hospital or community centre. These trade-offs in choice are conceptualised in basic models such as a PPB (production possibility boundary) where the trade-off between two commodities such as guns or butter can be visualised, and the optimum amount of production for both commodities can be demonstrated. Conceptual models such as the PPB are covered in Chapter 4, and this particular idea of a PPB can begin to demonstrate that choices have to be made and opportunity costs considered as there is a finite amount of goods and services attainable if growth is static.

Economies do expand and contract. This raises the possibility that more wants can be satisfied. But whether there is a limit to growth is the focus of Chapter 5. As discussed with respect to Malthus, environmental resource constraints may keep populations in check and constrain wealth. An exploration of what may restrain growth with reference to the built environment is important. For instance, if cities are the main hub and engine of employment and output, the ability for the city to grow economically may depend on its ability to house its workers or have available space to develop new industrial or commercial units. Key restraints to growth explored are those such as inadequate or inefficiently used natural resources, rapid population growth, inadequate human resources, cultural barriers, inadequate domestic savings, infrastructure and debt. Interesting thought experiments are brought forward such as asking what the consequences would be if growth was unconstrained and the free market was allowed to 'let rip'. An example could be to consider whether a complete relaxation of planning laws on green belts surrounding cities would add or subtract value over time.

d. Integrating theory of micro and macroeconomics

To further understand urban and environmental economics, the concept of markets in a microeconomic context is useful and is the principal focus of Chapter 6. As an introduction it considers what the economist Adam Smith (Smith, 1776) referred to as an 'invisible hand', where the forces of the producer and consumer interact, whilst tending towards an equilibrium of price and quantity for a particular market good or service, such as housing in a city. This particular system of production, consumption and distribution through market forces is the underlying mechanism for a predominantly market orientated economy where producers make a large number of decisions about what is produced. This is opposed to a more government-controlled decision-making system operating within a command economy. Extremes of complete free market and command economies are not polarised in reality and will tend to have a mix (i.e. a mixed economy), with leanings towards one particular model depending on its economic, historical and political context.

Free(er) markets are often cited as being more efficient in balancing the demand and supply of goods and services, although Chapter 7 will begin to demonstrate ways in which the market can fail. Market failure is seen when resources cannot be efficiently allocated between producers and consumers. For example, all of the costs generated by a coal power station may not be completely internalised if there is an external cost of pollution when producing energy. Study of externalities demonstrates the difficulty of valuing 'free' environmental resources that are used in the built environment.

Some social and environmental valuation techniques to be covered in Chapter 7 include: the socially optimal level of output, where the assimilative capacity of an externality is reached and therefore at an optimal level; and the willingness to pay (WTP) and willingness to accept (WTA) principles, which can be used to attach a value to economic and social goods that we may at first not be able directly to measure via the private market. As an example, the willingness to pay for the retention of recreation space could be met in part from car parking charges at the site (possibly including a tax contribution for a local authority), and this chargeable rate could be determined by balancing out what the visitors would be willing to pay and what they would be willing to accept as a charge. This rate could be applied as an average income per year and thus present some form of value. Obviously, this is a rather simplistic model with several issues that could be unpicked. The key point though, is that such social and environmental external costs and benefits can begin to be internalised if attempts at valuing them can be made.

Cost-benefit analysis (CBA) as another measurement tool is explained and discussed in Chapter 8. As an introduction, for CBA to be carried out, valuation needs to be attached to both the costs and benefits of a particular proposed project. In the example of building a new development on environmentally sensitive wetland, it will generate some social and environmental costs and benefits that will need valuation, for instance: the elimination of wildlife habitat costs; and benefits such as employment for those workers on the site. If costs and benefits are balanced against each other a net value can be placed on the development in addition to the more qualitative issues that surround its proposal – such as the sensitive subject matter of environmental protection. A further part of this valuation process is the use of discounting, which introduces an element of time, especially in that the real value of money will tend to reduce further into the future. This consideration is explained further in more detail in Chapter 8 to enable more accurate cost-benefit analysis given changes in the real value of money and the use of discounting.

Chapter 9 deals with macroeconomics, in the study of urban and environmental economics. Key macroeconomic subjects for study include the open economy, aggregate demand and supply, capital goods and flows, and globalisation. Widening discussion to the open economy enables study to open up to the whole economy, nationally and globally, and to understand its impact on urban and environmental concerns. The impact can be discussed in relation to an aggregated demand and supply (e.g. adding up all housing capital values), that is the sum of all demand and all supply for all the markets in a given spatial boundary (e.g. total housing capital value in a nation). Capital values in terms of financial capital can quickly and easily flow around the globe and will therefore be of relevance and interest to more macroeconomic study of the built environment, urban issues and environmental resources. This increase in flow and opening up of global foreign trade has been further enabled by the process of globalisation and will, too, be of importance in this text. For instance, the integration of capital financial markets has quickly enabled property investment around the globe to be quickly halted, or accelerated, based on globally accessed and shared market data.

Data and information at a macroeconomic level of study can be harnessed, analysed and used to influence decision-making. Chapter 10 centres on these considerations by providing understanding on macroeconomic government objectives and policy tools that influence urban spaces. Key macroeconomic objectives that require tools for intervention within macroeconomic policy include stable growth, stable prices, full employment, equilibrium in the balance of payments, protecting the environment, and the redistribution of income and wealth. The measurement of some of these objectives is also explained and discussed within Chapter 10 – particularly if there is going to be a need for appraisal, assessment and/or evaluation for macroeconomic government objective attainment.

Other economic forces operating at the macroeconomic level are covered in Chapter 10, such as exchange rates and interest rates. These forces can affect the macroeconomic government objectives (and be controlled) and hence need to be introduced with respect to urban areas and the resources that are produced and consumed. The demand for money to make transactions within a particular urban area will be influenced by the exchange rate for a particular nation. For instance, if commercial property is in high demand within a global city such as London, the demand for pounds to purchase or develop in that area will increase. The exchange rate is therefore indirectly connected to the country's level of business activity or gross domestic product (GDP). For interest rates, they are ultimately economic phenomena that relate to the trade-off between investment and saving. High interest rates would mean that there would be less borrowing and more saving, leading to less investment. This in turn would have implications for future growth if interest rates remained high. Urban development in property, for instance, would be stalled with high interest rates and therefore have a deflationary effect on the growth of the built environment.

With regards to managing the macro-economy, various government policies can be exercised as direct regulation or through monetary or fiscal policy. Regulation could be the direct imposition of a rule that stops urban development, such as the restriction of urban sprawl by the imposition of a green belt in planning law. Monetary policy can direct elements of how money is controlled through various mechanisms. Government policy could fix exchange rates to ensure certainty in money markets; it could adjust interest rates to effect investment and saving levels, or it could control the money supply by changing the number of notes and coins in circulation, therefore affecting the quantity and value of the currency. Fiscal policy is concerned mainly with taxation and spending levels by government in order to direct production and consumption. With regards to the built environment, the incentivisation of economic development in enterprise zones (EZ) through the lure of tax breaks is seen as one fiscal measure example; although the wider costs (public and private) of such urban development are debatable.

e. Applied economics of shared space and environmental resources

An understanding of whether the tools are being correctly applied can be realised in policy and practice. The first main element of applied economics is an introductory exploration of spatial models and concepts that have taken an economic element in their explanation (Chapter 11). Spatial models can draw on various city models, such as ribbon development, where housing has developed along ribbons of transport infrastructure such as roads, or urban development in the pattern of concentric rings, which formed on the earliest theoretical models to explain urban social structures. Demonstration of urban modelling will enable study in this text to connect with environmental resources that have enabled urban growth to expand, contract or be constrained. Issues of urban sprawl are particularly important to these urban developments and growth can be understood in relation to some urban areas being clustered in close proximity. This will involve ideas such as agglomeration and polycentric economies.

As well as conceptualising urban space and its connection to environmental resources, the study of specific sectors and land uses are important for understanding economic application. More in-depth study of the housing sector, for example, will bring forward urban economic issues in the supply of housing such as the rise in building carbon-neutral homes. Housing is also expressed in hedonic house-price methods that demonstrate price change over space (e.g. for a neighbourhood or city) in relation to other differentials in income, type, quality of property and the population.

Employment is an important applied urban economic issue to consider, as, for instance, 'wants' for housing (and its constituent environmental resources) will be partly determined by the supply and demand for labour in an urban area. Services available for urban populations also need economic analysis, as resource demand will be placed on institutions in the sectors of education, health and crime. Infrastructure can also be studied in more depth as an economic lens, as the connectivity and shared interests between places will need to be served. Infrastructure pressures could include issues of utilities and waste needs, in addition to transport connectivity between and within urban areas. Urban areas have difficulties, with a disproportionate amount of inequality and poverty, and economic discussion of such issues will be demonstrated. For instance, the stark contrasting inequality of wealth located within just a few streets or blocks in some of the major cities in the world warrants some serious critical thinking and analysis.

In continuing the theme of applied economics, Chapter 12 addresses what fundamental 'environmental resource' issues are of primary importance for their connection to the built environment. As discussed earlier, externalities require attention, as its effects are caused by a significant number of the population located within urban areas. Pollution is an obvious externality as urban areas expand and consume more products, and this can be captured and controlled using policy. The control and foresight into natural resource usage in energy production will have to be addressed, considering the continuing rate of urbanisation and subsequent expanding requirements of heat, power and light. These environmental aspects will have to be balanced against sustainable development objectives that also include economic and social priorities. For instance, the relatively cheaper extraction cost of coal or fossil fuels provides a quicker route to economic wealth for nations, but at a price in terms of environmental degradation or reductions in social well-being (e.g. health).

Valuing this resource will, in part, be formed by applied environmental economic methods that are now mainstream such as the use of environmental impact assessment (EIA) used in the majority of new major projects since the 1960s. International co-operation and agreement on environmental issues requires economic thinking at both a local and global level, especially as environmental problems have less regard for the national administrative boundaries they cross. Climate change is the most prominent of international and global concerns, and is receiving a great deal of attention. The economic consequences of climate change, such as extensive flooding and sea-level rise, will have economic consequences and may not necessarily be caused by the country being directly affected by climate change. In very basic terms, the economic cost of climate change will mean that it will remain on the international (and local) agenda for the foreseeable future.

f. The future and policy overview

As a conclusion to this introduction of urban and environmental economics, Chapter 13 will devote attention to how the concepts and their application are being addressed in policy given the contemporary challenges faced. The future of policy will depend on a decision to continue with current working policy, as well as adaptation and introduction of new regulation, guidance and incentives. Policies will be highlighted and discussed in relation to the built environment such as current and proposed regeneration (including economic development) strategies for urban areas. Environmental resource policy in relation to economic objectives will also be outlined and discussed. Furthermore, the overlapping and integration of urban and environmental policies will also be shown, especially if holistic and joint objectives can be realised. Part of this reaction and response will require careful detail, of which policy decisions can be made through the development of good information. Sophisticated economic evaluation and appraisal, for instance, can begin to close the loop on what have traditionally been opposing forces in the built environment of environmental protection and economic development. For instance, the most contemporary challenge that will be discussed (amongst others) is the need to balance social and economic justice whilst integrating a sustainable use of natural resources in a rapidly urbanising world.

Summary

1   Broad aims: (1) to unify thinking on urban economics and environmental economics; (2) to demonstrate what urban and environmental economics entails within theory, concepts and practice; (3) to provide an introduction of techniques and tools for research and further study.
2   Three interlocking strands that are viewed through an economic lens: urban issues; the built environment; environmental resources.
3   Particularly relevant to the planning, valuation and management of shared spaces
4   At the forefront of urban and environmental economics is the pressing issue that urban areas are set to become the focus in the global effort to curb climate change; especially for cities as contributors and as vulnerable spaces where its effects are realised.
5   Chapter overview with consideration of themes of: introducing urban and environmental economics; integrating theory of micro- and macroeconomics; applied economics of shared space and environmental resources; the future and policy overview.

Chapter 2

Perspectives in urban and environmental economics

Chicago, USA

This chapter looks at framing the concepts and theories that are considered within the study of urban and environmental economics. An initial definition and description is revealed in what is meant by 'urban' and 'built environment'. This is in addition to more 'natural' considerations of what is meant by the 'natural environment', 'environmentalism' and 'sustainability' for the purposes of study. Urban and environmental economic paradigms are demonstrated with particular reference to 'western' and 'ecological' views. Further consideration is given to the scale of activity when applying economic concepts to urban issues, the built environment and environmental resource allocation. Particularly as the spatial geographies of economic activity will generate different responses.

a. Urban *and* environmental economics

What is urban?

To understand what is meant by 'urban and environmental economics', the two constituent parts, 'urban' and 'environmental', need to be defined and situated in the cognate subject of economics. Reference to the urban is discussed in the disciplines of human geography and social science, as relating to phenomena in towns and cities. The traditional concept of a town or city is a freestanding built-up area with a service core with a sufficient number and variety of shops and services (ONS, 2001). A city also has administrative, commercial, educational, entertainment and other social and civic functions. With regards to connectivity, a city will generally have a local network of roads and other means of transport as a focus of the area, and it is also a place drawing people for services and employment from surrounding areas. For a town, its definition is in being a human settlement that is bigger than a village but smaller than a city. The size of a town is relative depending on the country in question as the geographic size of a town in the US, for instance, could be of equal size to a city in the UK. Often it is the administrative area that defines a town. For the UK it could be defined by whether it is administered by a borough or town council. However, sometimes, the boundary lies well beyond the town's built-up area and included tracts of rural countryside. More often, towns lie within the built-up area and so include only part of the totality of the urban area.

Cities have equal confusion in their definition and need further clarification and sub-categorisation to understand their make-up. Firstly, a city might simply be the historical core municipality (local authority area), such as the city of Chicago. Secondly, a city could be one of many cities (or municipalities) that make up a metropolitan area that can, too, be described as an urban area. For example, the Paris metropolitan area has 1,300 cities, the New York metropolitan area more than 700, and the St Louis metropolitan area nearly 400. Focus on urban analysis in this book is more on urban areas rather than metropolitan areas, the difference being that the metropolitan area tends to consider an entire labour market, whereas the urban area will describe an area of continuous urban development (or an agglomeration or urban footprint) that is almost never a single municipality. To make a further definition, an urban area can also take the form of a 'conurbation', when two or more urban areas grow together, as has occurred in Osaka-Kobe-Kyoto or the Washington and Baltimore urban areas, which are converging into single urban areas.

The urban area also contains key components of the central city, the urban core, and the suburbs. The central city or core city is viewed as the place (or metropolitan area) that emerged historically as the most prominent in the urban area. As an example, New York City is the prominent historical central city of the New York urban area. The urban core, as another element of the urban area, can contain the inner city and includes adjacent municipalities that developed during the same period as the core. Examples include, L'Hospitalet as a part of the urban core of the Barcelona urban area, and Cambridge as a part of the core of the Boston urban area core. The suburbs are also part of the urban area that is under analysis here and they are seen as part of the continuous urbanisation that extends beyond the core city. To illustrate with a London example, Epsom would be a suburb that is within the urban area as it is a municipality that is outside the Greater London Authority but inside the greenbelt. In considering these traditional components of a city, Chinese cities (as the word is translated) use slightly alternative layering of geography to denote their associated urban areas. These layers tend to take form conceptually at provincial level (e.g. Shanghai, Beijing), sub-provincial level or prefectural level. Prefectural levels are where there has been a provincial or autonomous regional subdivision. To highlight prefectural level connection to city status in China, out of 333 prefectural-level subdivisions, 283 are cities.

There are, however, at least three other approaches to defining an urban area. It may be defined either in terms of the built-up area (the bricks and mortar approach); or, alternatively, it may be defined in terms of the areas for which it provides services and facilities – the functional area. The functional area may embrace not only the built-up area but also freestanding settlements outside the urban area together with tracts of surrounding countryside if the population in these surrounding areas depends on the urban centre for services and employment. A third method is to use density (either of population or of buildings) as an indicator of urbanisation. For the UK, urban areas are defined with at least 20 hectares and at least 1,500 census residents. However, implementation of any of these approaches involves some arbitrary decisions in drawing up boundaries because, in practice, towns tend to merge physically and functionally with neighbouring towns and their hinterlands (ONS, 2001).

Urban areas, especially in England and Wales, can also be defined on the basis of land use that is irreversibly urban in character. For instance, the National Land Use Classification (DOE, 1975) defines 'urban land' as land that comprises several land use features. For instance, urban land is seen as containing:

(a) Permanent structures and the land on which they are situated (built-up site);
(b) Transportation corridors (such as roads, railways and canals), which have built-up sites on one or both sides, or which link up built-up sites that are less than 50 metres apart;
(c) Transportation features such as airport and operational airfields, railway yards, motorway service areas and car parks;
(d) Mine buildings (but mineral workings and quarries are excluded);
(e) Any area completely surrounded by built-up sites (areas such as playing fields and golf courses are excluded unless they are completely surrounded by built-up sites).

What is an environment?

The meaning of 'an environment' needs some short attention in order to frame the analysis of study being considered in this book. More broadly, the environment can be expressed as the external conditions and resources with which an organism interacts. More specifically, it is necessary here to consider differences between the natural environment and the built environment. The natural environment can simply refer to all living and non-living things that occur naturally on Earth, whereas the built environment refers more to constructed surroundings that provide the setting for human activity, ranging from large-scale civic surroundings to personal places. Beside the natural and built environments, the social environment can also be conceptualised as a sub-category, which considers the culture that an individual lives in and the people and institutions with whom they interact. In application to the approaches in this book, the natural environment will feature more as a part of natural resource usage, the built environment will link more with location in terms of an urban area, and the social environment will connect to urban issues that involve such individuals and institutions that interact with such resources in urban spaces.

b. The urban and built environment

It is important to note that there is a subtle difference between what is meant by urban areas and the built environment when exploring environmental and economic phenomena. As stated, an urban area relates more to the geographical location of towns and cities. This is in slight contrast to the built environment, which refers largely to all human-made surroundings. The built environment can therefore be constructed as a setting for several human functions such as housing, neighbourhoods, offices, factories, warehouses, transport hubs and networks, and supporting infrastructure such as energy, water and waste supplies.

An important point to note in framing this approach is that the built environment focus is within urban areas (i.e. predominantly towns and cities) rather than rural areas. Although interaction with the rural is no doubt excluded from thought due to its connectivity with urban areas – especially with regards to transport patterns such as rural commuters that work in urban areas (defined as within labour markets as the metropolis) – and accessing support infrastructure such as water reservoirs located in rural areas. This inter-connection between the urban and rural demonstrates that the two terms will not be mutually exclusive. However, considerations within this book's approach will be to focus on urban areas and the issues contained within them, whilst being mindful of and drawing in rural connectivity, its use and consumption of resources in the natural environment, and as a consequence its integration with environmentalism and environmental sustainability.

In practice, the term 'built environment' is used to encompass several inter-disciplinary professions to which it relates. Built environment professions include architecture, planning, landscaping, construction, design, management, urban planning, urban design, housing and regeneration. The built environment as a focus for these professions is not traditional in the academic sense, and built environment ideas often draw from long-established academic disciplines to provide and apply more conceptual and theoretical work. As such, more established disciplines that can be drawn on are those such as geography, environmental science, law, public policy, management, design and technology. More importantly, in the case of this book, the built environment will connect to interlocking strands of urban issues and environmental resources – whilst engaging with disciplinary ideas of economics (and sub-disciplines such as environmental economics and urban economics) and situated with emphasis on shared urban spaces.

c. The natural environment, environmentalism and sustainability

For urban and environmental economics the natural environment is critical to an urban area's functioning, prosperity, well-being and health. The natural environment encompasses all living species and non-living things that have occurred naturally. With further specificity, Natural England (2008) views the natural environment as meaning our landscapes, air, flora and fauna, freshwater and marine environments, geology and soils. The natural environment can be sub-categorised into two components of either universal natural resources or ecological units. Universal natural resources are those such as air, water and the climate, which does not have clear boundaries, as well as non-human created resources such as energy, radiation, electrical charge and magnetism. Ecological units as a component of the natural environment are those functioning natural systems that operate without much human intervention. These ecological units, which can systemically function within a particular boundary, are those such as vegetation, soil, micro-organisms, atmosphere and rocks. With respect to urban areas, both the consumption of natural resources and the healthy ecological functioning of the natural environment will depend largely on a sustainable approach that has been at the forefront of environmentalism dating back to the nineteenth century in modern western thought.

Environmentalism in social science terms is a political and ethical movement that strives to protect and improve the quality of the natural environment by human activity. In doing so, environmentalism takes on a philosophical stance that living things in nature deserve a degree of moral reasoning and consideration that are manifest in political, economic and social policy. Environmentalists therefore act as an individual or organisational voice to promote environmentalism and influence the political process through lobbying, activism (e.g. grass roots campaigns and protests) and education. Promotion of environmentalism is particularly important for urban areas and is not confined to rural areas, which by definition have more of a 'natural' environment. Urban areas consume vast amounts of natural resources and therefore will have pressure from the environmental movement to (a) maintain and protect the supply of natural resources and quality of ecology that are provided in urban areas; and (b) affect the consumption of natural resources and quality of ecology by those people that function in urban areas – consumption could be by those that reside, work or briefly visit a particular area.

Institutionally, the promotion of environmentalism has been relatively new in terms of its recent inclusion in school curricula. Environmentalism has also matured as a political institution with several primary foci for its permanent place in social conscience. These include: Environmental Science, Environmental Activism, Environmental Advocacy, and Environmental Justice. However, it should be noted that environmentalism has existed in some form for centuries. Environmental protection in the past has centred on pollution control; for instance, the earliest known writings that concerned environmental pollution were Arabic medical treatises written during the Arab agricultural revolution (700–1100) by writers such as Alkindus (Worster, 1977). Moreover, pollution control can be seen when King Edward I of England banned the burning of sea coal in 1272 after its smoke had become a problem (Kovarik, 2011). Historical environmental concerns such as these tended to centre on air contamination, water contamination, soil contamination and solid waste mishandling. Air pollution would become the dominant problem during the industrial revolution from the eighteenth to the nineteenth centuries, and into the more recent past of the twentieth century with the great smog in London in 1952. More contemporary environmental focus is the growing concern of CO_2 release and subsequent issue of global warming and climate change.

Concern over climate change and a focus to improve and protect natural resources and improve ecological health have been promoted via the (environmental) sustainability agenda. Sustainability, in simple terms, is the capacity to endure. In more contemporary writing, with regards to economics, 'sustainability' has been described as an economy in equilibrium with basic ecological support systems (Stivers, 1976). Sustainability has since taken on greater meaning with regards to the environment and its connectivity to development. In 1987, the World Commission on Environment and Development developed a definition of sustainability that was included in its findings, which became known as the Brundtland report (WCED, 1987). The report stated that sustainable development meets the needs of the present without compromising the ability of future generations to meet their own needs (ibid.).

Although the Bruntland report definition has become widely publicised, the term 'sustainable development' is not limited to one precise definition, although much debate tends to centre on three pillars of sustainable development, which consider economic sustainability, social sustainability and environmental sustainability. This also has an underlying premise that sustainable development ties together concern for the carrying capacity of natural systems with the social challenges facing humanity. Important for the study of urban and environmental economics is that sustainability interfaces with economics through the social and ecological consequences of economic activity. As such, the approach taken here ties intellectually with ideas of 'sustainability economics', which involves ecological economics where social, cultural, health-related and monetary/financial aspects are integrated into the study of urban areas.

d. Modern western and ecological paradigms

Perspectives on urban and environmental economics can be taken from two distinctive types that are either more 'modern western' or 'ecological' in thought. Modern western thinking often draws on environmental resource economics (ERE), which draws back to largely classical economic ideas and their subsequent development in explaining contemporary issues. For instance, a modern western economic idea is that of externalities where a cost (that potentially could be valued) is incurred by a third party external to the market transaction between the buyer and seller. Within an urban setting, a negative externality could, for example, be the cost of pollution incurred by residents (the third party) due to the operation of heavy industry (a producer of goods for consumption) in proximity to a neighbourhood. In this instance, western economic thinking will attempt to value this external cost and then internalise it within a market framework to realise the full cost of the heavy industrial operations.

Alternatively, ecological economic (EE) paradigms tend to integrate biophysical perspectives into the interaction between the environment and the economy. Ecological economic thinking, for instance, would be integrated into a range of natural and social sciences such as thermodynamics in the case of energy production, and ethics in the case of actions on all living things, rather than a single focus on human social interaction. The ecological economic approach also brings with it a significant amount of connection to resources – natural and human, tangible and intangible. The traditional environmental resource approach by definition focuses on resources but with less of a regard to ecology. In connecting to a resource approach, ecological economics in an urban setting would, for instance, look at how urban areas produce and consume environmental resources (e.g. energy, food, building materials) in relation to the resource's ecological protection, improvement or degradation. Consideration of ecology as a system with the ability to reproduce resources enables part of a much-neglected physical side of the economy. Furthermore, this incorporation of ecology also allows economic analysis at differing scales, such as at the individual, household, neighbourhood, city, region, metropolis and global levels. The ecological ability to grow higher-value healthy food in one whole urban area (i.e. town or city) could, for instance, be contrasted to inequalities of neighbourhood access to quality small spaces to grow food.

The constituent components of both modern western 'environmental resource economic' (ERE) and ecological economic (EE) approaches are displayed in Table 2.1 in order to compare and contrast their view more clearly. Traditional environmental economics is concerned with the allocation of scarce environmental resources, and if they are allocated at a cost to a third party (e.g. to pay for pollution), an externality is seen to exist. This is in contrast to an EE concern of whether a resource will be exhausted at a particular scale of geography. ERE tends to deal more with efficiency ideas, especially those of internalising external welfare costs into the whole-market system, or providing explanation in terms of Pareto efficiency where no further improvements in allocation can be made to make one person better off without making another worse off. To be Pareto-efficient is therefore a situation where in making any further recipients better off, at least one person will be worse off. The alternative is an inefficient situation where both individuals and groups can be made better off by the additional resource, primarily because the resources being used have not reached their limits. As a simple example in urban areas, a Pareto-efficient allocation of housing is if there is an optimum allocation of housing at an acceptable quality of shelter, and any further allocation would begin to make at least one individual worse off. In essence, ERE in using Pareto efficiency can only explain how efficient an allocation is (a point where there is minimal waste – such as in the efficient provision of housing at a certain quality), not how the distribution is made in terms of equity or well-being. In EE, less of a focus on efficiency of resources is made due to the potential for regenerative processes in producing and assimilating environmental resources – the optimal efficiency point is more fuzzy and limitless (as far as its reproductive capacity) is concerned.

With regards to growth, ERE and EE hold ideas of sustainability, although ERE views this growth optimistically in terms of what gains can be made, rather than in terms of what difficult choices will have to be made to ensure development for inequitable spaces. For instance an ERE approach to food consumption in developed urban areas would see the competitive advantage for farmers in less developed countries as they supply produce to them. However, EE would take a more pessimistic view of this food consumption in developed urban areas supplied by less developed countries in that the sustainable development of the producer countries would be stifled if focus on agri-business was the only economic choice and option for growth. Deterministic optimisation of inter-temporal welfare is seen as a key feature of ERE where welfare is more rationally optimised at different timescales (e.g. see Pareto efficiency approach). This is in contrast to EE thinking, which is less rational and linear; for example, welfare would evolve (or devolve) at a rate that is less rigid or unpredictable.

Differences in temporal considerations in ERE and EE are apparent. For instance, EE takes a more short- to medium-term focus, as, for instance, the extraction or use of a resource is taken at a present value, with some element of discounting (to account for changes in the value of money), into the future, say 25 years. EE takes a longer-term focus engaging with ideas that certain resources have a useful (economic) life, which could continue into the future indefinitely, especially if taking a more ecological approach, where resources have reproductive capacity. This difference and the examples chosen also demonstrate that ERE has an emphasis on abstract concepts such as monetary indicators that can generate external costs and values, whereas for EE it tends to have emphasis more on biological and physical indicators that operate within a system.

For ERE, its content considers resources as a small constituent part of a wider picture, which can be analysed within a single-discipline lens such as economics; for instance, it draws on classical economic theory such as maximisation, utility (a measure of satisfaction) and profit. In contrast to this abstract theory, EE would engage in more contemporary behavioural economic approaches, such as analysing choices, and the rational and irrational in those choices, made by those involved in the process. This involvement of individual choice and decisions by agents also highlights that EE tends to consider involvement of agency, rather than focusing on the operation of a global market without individual actions that occur in connection with market change. As a final contrast, it can be seen (Table 2.1) that this greater inclusion of agency in EE enables more thought for environmental ethics and less of a functionalist modeled approach in ERE where for instance the quantity of consumption of a resource in an urban area could be equated to an amount of utility (satisfaction) experienced by those that consume the resource.

Table 2.1 Differences in economic emphasis between traditional environmental resource economics (ERE) and ecological economics (EE)

Traditional environmental and resource economics	Ecological economics
1. Optimal allocation and externalities	1. Optimal scale
2. Priority to efficiency	2. Priority to sustainability
3. Optimal welfare or Pareto efficiency	3. Needs fulfilled and equitable distribution
4. Sustainable growth in abstract models	4. Sustainable development, globally and north/south
5. Growth optimism and 'win–win' options	5. Growth pessimism and difficult choices
6. Deterministic optimisation of inter-temporal welfare	6. Unpredictable co-evolution
7. Short- to medium-term focus	7. Long-term focus
8. Partial, monodisciplinary and analytical	8. Complete, integrative and descriptive
9. Abstract and general	9. Concrete and specific
10. Monetary indicators	10. Physical and biological indicators
11. External costs and economic valuation	11. Systems analysis
12. Cost-benefit analysis	12. Multidimensional evaluation
13. Applied general equilibrium models with external costs	13. Integrated models with cause–effect relationships
14. Maximisation of utility or profit	14. Bounded individual rationality and uncertainty
15. Global market and isolated individuals	15. Local communities
16. Utilitarianism and functionalism	16. Environmental ethics

Source: Adapted from Van den Bergh (2000)

More philosophically (and rather simplistically), the traditional and ecological approaches have characteristics that transcend some of these economic distinctions and are worth attention to further frame these alternate paradigms. Pearce *et al.* (2000) in Table 2.2 highlights that the traditional western view holds that there is domination over nature where the human is supreme, and that nature is seen as a resource and as an input into a system. This is in contrast to the ecological view that would have more of a harmony with nature, and is more egalitarian; plus it takes a more biocentric view that nature is valued for what it is rather than what value it has in monetary terms. Both viewpoints see that things in the natural world are tangible and could (rightly or wrongly) be commodified. As a result, natural tangible commodities are contrasted in that they could more traditionally be viewed as being ample, as the function that a good provides (e.g. food for human consumption and life) is ample as far as the capacity of the system allows (e.g. fresh water captured and stored more locally – such as tap water from a nearby reservoir). And if a particular commodity reaches its finite supply (and subsequent increase in price), a switch by consumers to other substitutes will occur (e.g. more expensive transported bottled fresh water). A more ecological view would see that commodities are finite in their physical being, and thus they are consumed or degraded until the point that they do not exist or can be used in their physical form any more (e.g. no available fresh water).

Contrasting lifestyle characteristics are also distinguished between a simple western and ecological view. A western view is more consumerist and competitive in lifestyle and has an overarching objective to maintain economic growth. This is in contrast to a more ecological view, which is more co-operative and has less of a connection to material goods. This is by focusing on developing knowledge and self-realisation as well as 'making do' with what a person has or by recycling. Politically, the two distinctive views are seen to differ starkly, with the western view being more hierarchically controlled at the national centre. This is different to an ecological political view that centres on a more non-hierarchical and decentralised (from regional to neighbourhood, or even in smaller scale from household to individual) scale with a grassroots democracy emphasis on biological concerns.

As a final contrast of the philosophical western and ecological views on the environment, the modern western view concentrates on high technology such as focusing on technologically advanced energy production in any form, whereas the ecological view would centre attention on the most appropriate form of technology such as using renewable energy sources in urban areas in order to protect the environment. In connecting these broader views on the environment, it can be deduced that the modern western view draws closer ties to ERE economic approaches, and the ecological view tends toward more EE approaches. The boundaries between both views and economic approaches are obviously fuzzier, and fluidity of components will not contain them to a particular category. However, movements in the sustainable development agenda have enabled more ecological views to synthesise with more traditional western economic thinking – especially with the growing study across many academic institutions of environmental economics.

Table 2.2 Western and ecological views on the environment

Traditional western view (modern)	Ecological view
Domination over nature	Harmony with nature
Nature as a resource, an input	Nature is valued for itself
Human supremacy	Biocentric egalitarianism
Ample resources/substitutes	Supplies limited
Economic growth is an important objective	Non-material goods, knowledge/self-realisation
Consumerism	Doing with enough/recycling
Competitive lifestyle	Co-operative lifestyle
Centralised/urban/national focus	Decentralised/bio-regional/neighbourhood focus
Hierarchical power structure	Non-hierarchical/grassroots democracy
High technology	Appropriate technology

Source: Adapted from Pearce *et al.* (2000)

e. Scale: global, regional and local approaches

The scale of activity when looking at different perspectives in urban and environmental economics is important. It was highlighted previously that incorporation of ecology into economic thinking could enable analysis at differing scales such as at the individual, household, neighbourhood, city, regional, metropolis and global levels. It was also noted using the example of food resources in urban areas that these different scalar approaches could demonstrate that micro-scale study (such as the neighbourhood) could be both aggregated up to a global-scale study, but also be shown to have differing dynamics operating at each scale. Dynamics at a local scale enable a more pragmatic and practical scope, for instance urban and environmental economics could study the poor socioeconomic circumstances of a disadvantaged neighbourhood that reinforce a poor local environment. At a global scale, other dynamics could be considered, as the boundary of analysis is literally 'outer space', meaning that broader and far-reaching questions and analysis can be tackled. If the Earth is conceived as a single unitary organism, study can more comprehensively explore and suggest whether there are limits to growth using ecologically sensitive resources to meet urban needs, especially if it can be argued at a global scale that the ecological reproductive capacity of the Earth is finite.

Summary

1   An 'urban' area refers to a town or city as a free-standing built-up area with a service core with a sufficient number and variety of shops and services. A city also has administrative, commercial, educational, entertainment and other social and civic functions.

2   With regards to connectivity, a city will generally have a local network of roads and other means of transport as a focus of the area, and it is also a place drawing people for services and employment from surrounding areas.

3   Often it is the administrative area that defines an urban area. A city might simply be the historical core municipality (local authority area) or could be one of many cities (or municipalities) that make up a metropolitan area that can, too, be described as an urban area.

4   Focus on urban analysis in this book is more on urban areas rather than metropolitan areas, the difference being that the metropolitan area tends to consider an entire labour market, whereas the urban area will describe an area of continuous urban development.

5   An urban area can also take the form of 'conurbation' when two or more urban areas grow together and converge into a single urban area.

6   The urban area also contains key components of the central city, the urban core and the suburbs. Chinese cities tend to take form conceptually at provincial level (e.g. Shanghai, Beijing), sub-provincial level or prefectural level.

7   An urban area may also be defined as: (1) the built-up area (the bricks-and-mortar approach); (2) in terms of the areas for which it provides services and facilities – the functional area; (3) high density of population or of buildings as an indicator of urbanisation; or (4) land use that is irreversibly urban in character.

8   The natural environment can simply refer to all living and non-living things that occur naturally on Earth, whereas the built environment refers more to constructed surroundings that provide the setting for human activity.

9   The social environment can also be conceptualised as a sub-category, which considers the culture that an individual lives in and the people and institutions with whom they interact.

10 The built environment is constructed as a setting for several human functions such as housing, neighbourhoods, offices, factories, warehouses, transport hubs and networks, and supporting infrastructure such as energy, water and waste supplies.

11 In practice, the term 'built environment' is used to encompass several interdisciplinary professions such as architecture, planning, landscaping, construction, design, management, urban planning, urban design, housing and regeneration.

12 The built environment ideas often draw from long-established academic disciplines such as geography, environmental science, law, public policy, management, design, technology and economics.

13 There is a need to be mindful of rural connectivity in the use and consumption of resources of the natural environment and integration with environmentalism and environmental sustainability.

14 The natural environment is critical to an urban area's functioning, prosperity, well-being and health. It refers to our landscapes, air, flora and fauna, freshwater and marine environments, geology and soils. The natural environment can be

sub-categorised into two components of either universal natural resources or ecological units.

15 Promotion of environmentalism is particularly important for urban areas and is not confined to rural areas that, by definition, have more of a natural environment.

16 Concern over climate change and a focus to improve and protect natural resources and improve ecological health have been promoted via the sustainability agenda.

17 Modern western thinking often draws on environmental resource economics (ERE). Ecological economic (EE) paradigms tend to integrate biophysical perspectives into the interaction between the environment and the economy.

18 Incorporation of ecology into economic thinking can enable analysis at differing geographical scales such as individual, household, neighbourhood, city, region, metropolis and global.

Chapter 3

The built and natural environment according to economists

Amsterdam, Holland

Ideas, concepts and theories are developed from intellectual thoughts that are of importance when studying urban and environmental economics. This chapter will provide initial thought on the origins of economic thinking that underpin its application to urban areas and environmental resource allocation. 'Modern' economic thought will be expressed from the writings of significant and pivotal thinkers such as Thomas Malthus on population, David Ricardo on trade and rent, and John Stuart Mill on limits to growth. As the economic discipline has progressed, so too has its applied focus to urban and environmental issues. Of note here is the work developed further by Arthur Cecil Pigou on welfare economics and Ronald Coase on behaviour. Outline is also provided on the contribution from the Club of Rome as an institutional force. Contemporary urban and environmental issues are introduced with regard to explaining what key challenges are prevalent, such as urban heat islands, climate change and global warming, and sustainable development. Attitudinal perspectives on environmental allocation with respect to urban space are concluded in the chapter.

a. The origins of economic thinking

The origins of economic thinking can be conceived to date back to when humans bartered for goods or exchanged through a system of gifts. From 9,000–6,000 BC, livestock was often used as a unit of exchange as barter. Later, as agriculture developed, people used crops for barter. For example, a farmer could ask another farmer to trade a pound of wheat for a pound of fruit. Aside from the obvious issue of fair trade, barter has other problems such as timing restraints and quantity restrictions. If you wish to trade wheat for fruit you can only do this if the wheat is needed in the near future or if the fruit is in season or available locally. This is where money as a commodity becomes part of the economic system; where, in particular, an 'IOU' or intermediary resource is needed to counter time delays and quantity restrictions of obtaining a good. This intermediate commodity can then be used to buy goods, such as fruit, when they are needed. Thus the use of money makes all commodities become more liquid or easier to get hold of – cash is a very liquid form of money today.

When such an intermediary is introduced, this becomes the basis of a commodity currency – money backed by a multilateral barter agreement between all participants. Historic origins and examples of such commodity currency include cowry shells in 1200 BC in China, which subsequently developed at the end of the Stone Age into a system of mock cowry shells made with holes in so that they could be threaded into chains. The use of metal coins as commodity currency originates from 500 BC where silver was used imprinted with various gods and emperors to denote value. The system of commodity currency in many instances evolved into a system of representative currency. This occurred because banks that came into being to store individual and group wealth would issue a paper receipt to their depositors, indicating that the receipt was redeemable for whatever precious goods were being stored (usually gold or silver money). This allowed receipts to be traded as money (known as bonds), and for much of the nineteenth and twentieth centuries many currencies were based on representative money through use of the gold standard.

Contemporary western systems of money are traded digitally as numbers on a computer screen, and wealth can be symbolised by the differences in ones and zeros held in an account. Systems of exchange are not succeeded in a linear timeframe as human trade will take whichever mode suits its business objectives. For instance, bartering is still part of the ICT (information communication technology) age, with web technology being utilised to open up a set of complex business models on the basis of barter. It is estimated that the worldwide organised barter exchange and trade industry has grown to an $8 billion-a-year industry and is used by thousands of businesses and individuals.

This introduction to the development of tools and methods in economics (e.g. barter, gifts, trade, commodity currency, bonds, digital exchange) sets the historical context to the study of how economics can be applied to environmental resources and the built environment with a focus on urban space. The forces that operate to distribute resources and wealth can shape the development of people and place – and this force has not changed since the origins of human exchange. What is important for this chapter is to understand how these economic forces have been perceived to influence built and natural environments in the context of a historical shift from rural to urban growth at a global scale. Much of this thinking in western thought has its origins in the Enlightenment (or age of reason) from approximately 1650 to 1800, where an elite of intellectuals promoted intellectual interchange and opposed intolerance and abuses in Church and state. Towards the end of this period (eighteenth and nineteenth centuries) and entering a philosophical romantic period (second half of the eighteenth century), significant thinkers lived within the industrial revolution and this shaped their writing on how society was adapting to industrial development. Romanticism was harking back to a period of nature as industry grew, population rose and urbanisation spread, while any fears of these changes were juxtaposed against industrial benefits such as average incomes rising and population beginning to exhibit unprecedented sustained growth.

b. Thomas Malthus

As introduced in Chapter 1, Thomas Malthus (1766–1834) was one writer who took concerns over economic growth and population rise as a consequence of the industrial revolution and extrapolated via thought experiment as to how this population growth would affect social improvement with respect to environmental (natural) protection (Malthus, 1798).

The Malthusian problem (or similarly phrased Malthusian catastrophe, Malthusian check, Malthusian crisis, Malthusian disaster, Malthusian fallacy, Malthusian nightmare or Malthusian theory of population) that still resonates in contemporary thinking is the question of whether per capita incomes and consumption would inevitably be driven down due to the rise in population levels. This concept is held in check due to the expansion of territories and falling birth rates in more developed countries (such as the US and western Europe).

Statistically, it was argued during the time of Malthus's writing that a nation's population had a tendency to double every 25 years. This meant that, mathematically, its population grew in a geometric pattern by periodically doubling itself – say from 2 million, to 4 million, to 8 million, to 16 million etc. This pattern could be set against a nation's food supply that would have to support this population growth. It was argued that food supply growth for consumption increased at a more uniform periodic arithmetic rate as more land was used to produce food –

say 2 million tonnes of corn, to 4 million, to 6 million, to 8 million etc. These differences in the reproductive capacity of humans and the reproduction of food available to sustain them were seen to diverge and thus place an ultimate check through food shortage on population growth (Figure 3.1). Socioeconomic impacts of this check were also seen to affect poorer inhabitants of society in urban areas as food became scarce. For instance, higher infant mortality in urban areas through lack of nutrition within more disadvantaged groups is recognised when Malthus states that 'this mortality among the children of the poor has been constantly taken notice of in all towns' (ibid., 1798, p. 23).

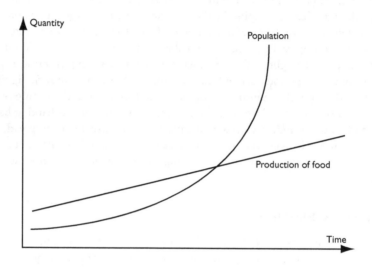

Figure 3.1 The Malthusian problem of population growth and food supply
Source: Author

The simple duality of population growth and food availability is recognised, in the Malthusian problem, and there are many critiques that can be attached to a more complex issue. For instance, advances in technology may improve the productive capacity of food as seen via developments in fertiliser or pesticides as part of the green revolution. Technological developments or education in birth control may also adjust the dynamics of population increase. This is in addition to wider attitudinal changes such as voluntary restrictions of family size in western Europe, or more direct regulatory forces such as the one-child policy in China that, in principle, restricts married urban couples to having only one child (with exceptions such as ethnic minorities and parents without any siblings themselves).

c. David Ricardo

To begin to broaden this economic thinking on environmental resources in urban areas, the work by David Ricardo (1772–1823) provides an additional dimension of how trade and relative price will affect growth. In his book, *The Principles of Political Economy and Taxation*, Ricardo (1817) dissects how economic concepts such as rents (and profits and wages) were applied to theories of land capacity and population. Rents are viewed as the difference between the produce obtained by the employment of two equal quantities of capital and labour. So if land could be used for either cultivation (a price and quantity willing to be paid by a labourer) or for residential purposes (a price and quantity willing to be paid by the household), the rent charged would be the difference in these different land-use options. At the forefront of Ricardo's thinking was a conclusion that land rent (the reward for owning land) grows as population increases. This would mean that with population growth increasingly concentrating in urban areas, consequences would involve a rise in rents of real-estate values and a rise in rents of land used. This would generate over a period of time an aggregate appreciation in the value of property and necessity goods such as food.

Ricardo was also instrumental in developing the concept of comparative advantage, where at a national scale, trade could increase economic growth as long as each nation had different relative costs for producing the same goods. He argued for competitive advantage as all nations could benefit from free trade, even if a nation was less efficient at producing all kinds of goods than its trading partners. This could mean that if a developing nation had an abundance of low-cost materials for manufacturing goods (e.g. silicon), it could trade on this competitive advantage in order to re-balance its other resources (land, labour, capital, entrepreneurship) that need appropriation or greater efficiency for economic development.

Competitive advantage theory has counter-weight problems in that the type of advantage chosen will form to meet the needs of the rich and powerful and not be a bilateral exchange. For instance, the type of competitive advantage chosen, such as a nation concentrating on silicon, may eventually have development gains (and adding value) by manufacturing silicon components rather than just extraction of silicon compounds. However, wealthier nations will be in a stronger trading position to manufacture and trade silicon parts within its own higher added value service sector. This theory is further complicated in that different resources for competitive advantage will be desired during different phases of the development process. For instance, the trading advantages from steel during a global age of heavy manufacturing will be different to the trading advantages of silicon during a digital revolution. The rules of trading and relative prices of commodities will be very different and in greater control of the more developed nations than in previous decades, especially if a greater inequality exists between the developed and developing nations. What is referred to as a Ricardian vice may ring true in that rigorous logic does not always provide a good economic theory, and as a result, alternative theory must be sought.

d. John Stuart Mill

John Stuart Mill (1806–1873), in his book on the principles of political economy (1848), further developed economic thinking that drew on ideas by Ricardo and the significant neoclassical economic forefather Adam Smith (Smith, 1776). The book by Mill became a leading economic textbook for 40 years after it was written as it condensed key microeconomic and macroeconomic theories whilst contributing more refined and advanced ideas on competitive advantage, economies of scale and opportunity cost. Mill (1848) significantly expressed the analysis of changes in population growth (particularly in urban areas) and changes in productive capacity of necessity and luxury goods. He argues that there were limits to such population growth globally, and by extraction in urban areas, because he viewed that growth in the economy and growth in nature were not endless, unbounded processes and that they must eventually return to some lasting equilibrium. Linear economic growth was seen due to humanity's struggle (especially with respect to western societies) to obtain material advance during the eighteenth, nineteenth and twentieth centuries. In Book IV, Chapter VI, 'The Stationery State', Mill takes the view that this material advancement may be to the detriment of 'real' wealth in terms of quality of life for individuals and society if the environment is being degraded too severely. With regard to concerns over the environment and quality of life from unlimited economic material gains, he states that:

> If the earth must lose that great portion of its pleasantness which it owes to things that the unlimited increase of wealth and population would extirpate from it, for the mere purpose of enabling it to support a larger, but not a better or a happier population, I sincerely hope, for the sake of posterity, that they will be content to be stationary, long before necessity compel them to it.
>
> (Mill, 1848)

This thinking begins to bring in ideas of social welfare as both environmental and social concerns are attached to the economic. For instance, there is thought that well-being or quality of life can be maximised when all individuals in society can all consume at the greatest capacity. Well-being is not thought of as an 'average' or 'mean' position where inequalities of wealth distribution could skew the view of well-being for the majority of the population. Movement towards ideas on social drivers (including concerns for the environment as it interacts with the population) became more prominent in the latter writings of Mill as he considered the economy in combination with social forces as well as how economics can be thought of as aligned to all things environmental.

The abolition of inheritance tax and the development of a co-operative wage system were two significant social leanings towards the end of Mill's writing. This considered how a significantly growing number of properties were owned and distributed in urban areas, plus for an increasingly urbanising labour force the distribution of wealth and power as well as environmental degredation. This was particularly manifest in the rise of union power by those working and living near to labour-intensive manufacturing industries during the industrial revolution.

e. Other significant authors of economics and natural resources

As Mill opened the way for neoclassical thinking in social progress, Arthur Cecil Pigou (1877–1959) brought welfare economic thinking to the fore. As part of the neoclassical writers on welfare economics, Pigou took Alfred Marshall's (1842–1924) *Principles of Economics* (1890) and concept of externalities and embedded it further into the discipline. Externalities, as introduced in Chapters 1 and 2, are those benefits gained or costs incurred by a third party due to the transactions of buyers and sellers of a good or service. Therefore there can be positive or negative effects on individuals or groups that are external to the market – hence the term 'externalities'. These third-party effects give rise to the notion that a third force (not just producers and consumers) should be able to step in and correct any external benefits or costs. Examples of externalities can link in with urban issues where, for instance, incidents of health problems by residents near to factories will provide a negative externality cost. This is certainly the case for Chinese cities such as Shanghai where pollution levels (from industry and transport) provide breathing difficulties for residents that may not gain any direct market benefit from the economic activity taking place.

As well as externality examples as urban issues, the development of the built environment can also draw out some cases. For instance, as a simple site-specific example, a new-build property will have to consider the light that may be taken away from existing residential properties, as in the legal requirement of a 'right to light'. If the existing residential property does not gain financially from the market transaction of the adjacent new-build property, but does experience a cost in the amount of light exposure, the residents will be experiencing some degree of negative externality. A positive externality example could, for instance, be that a resident paints the exterior of his/her house, and therefore raises the aesthetic and economic exchange value of the property, but also raises the third-party value of the surrounding neighbourhood and the wider community that benefit from the space in view of the freshly painted house. It should be noted that externality concepts need careful analysis as benefits and costs may be valued differently. For instance, in the house-painting example, a house painted without care or attention in keeping with what is socially acceptable (e.g. the house could be painted black!) could generate a negative rather than a positive externality for the wider community.

In order to deal with externalities, Pigou believed that taxes could provide a disincentive to negative externality (or internalise the cost of) problems such as pollution, or that subsidies could provide an incentive to provide (or internalise the cost of providing) positive externalities for the wider society external to the market. The use of taxes or subsidies with regards to dealing with externalities is still referred to in economic circles as Pigovian taxes and subsidies. A fundamental difficulty with this neoclassical approach is the fundamental belief that markets are central – more modern progressive economic thinkers have held that the market may not be central to analysis of phenomena, particularly for the urban and environmental question, as markets can indeed fail.

Market failure is, in essence, where the market does not efficiently allocate all goods and services. For instance, a large concentration of empty homes lying idle until their eventual decay would appear to be a wider inefficient use of resources. The detail of market failure will be covered in Chapter 6, but what is important here with regards to economic thought is that public-choice economists such as Ronald Coase in the 1960s would begin to demonstrate that behaviours and actions outside of the invisible hand of the market, and outside of the incentives of tax and subsidy, could affect economic activity. Of particular significance was that integration of legal considerations will affect the decision to carry out a particular economic activity. For example, bargaining over land due to legal costs between parties may generate different transaction costs irrespective of what the open-market value suggests (Coase, 1960).

f. Club of Rome

It can be seen that classical economic thought on the urban and the environment is relatively young with writers dating back to the eighteenth century. This is not to suggest that urban settlements did not have pressing social and environmental considerations prior to their intellectual conception. For instance, problems of sanitation, pollution and availability of resources in urbanising towns are documented prior to the formalisation of intellectual debate. Furthermore, there would have been interest groups beyond scholarly writers that articulated urban and environmental difficulties that would need to be overcome in order to further their own individual and collective interests. For instance, political parties, informal pressure groups, social movements, land and property owners, and merchants held urban and environmental interests. In the more recent past, groups of both scholars and wider interest groups have begun to propel urban concerns and environmental problems that face humanity.

An iconic grouping of individuals that gathered in 1968 to address such problems were those that gathered under what was to be known as the Club of Rome (2009). This organisation formed as a small international group of professionals from the fields of diplomacy, industry, academia and civil society. It is quoted that their particular concern was the long-term concerns regarding unlimited resource consumption in an increasingly interdependent world (Club of Rome, 2009). The most significant report that launched the group globally was the *Limits to Growth* report (Meadows *et al.*, 1972) commissioned by the Club from a group of systems scientists at the Massachusetts Institute of Technology. The report explored a number of scenarios and stressed the choices open to society to reconcile sustainable progress within environmental constraints. The normative approach to this subject was clear and stressed the importance of such attention on a world stage; the report for instance questioned:

Is the future of the world system bound to be growth and then collapse into a dismal, depleted existence? Only if we make the assumption that our present way of doing things will not change. There is ample evidence of mankind's ingenuity and social flexibility...Man must explore himself – his goals and values – as much as the world he seeks to change. The dedication to both tasks must be unending. The crux of the matter is not whether the human species will survive, but even more, whether it can survive without falling into a state of worthless existence.

(Pearce *et al.*, 2000)

This apocalyptic vision of the future without attention carried great weight and has provided programmes such as 'A New Path for World Development' (Club of Rome, 2009). Most importantly, the Club led the way for inter-disciplinary and inter-institutional thinking and subsequent action on environmental issues. As well as resource depletion, the Club began to enable progress into additional contemporary concerns of global warming, climate change and sustainable development. These themes are the new challenges facing leading academics and practitioners, especially for those focusing on urban areas that have become spaces where these issues tend to be produced and consumed.

g. Contemporary urban issues and the built environment

Since the Club of Rome, contemporary challenges to urban and environmental economics have been set at different spatial scales from the global to the local (to consider neighbourhoods, households and individuals). At a global scale, global warming and climate change have dominated thinking in relation to urban areas and environmental resources. At a more basic urban analysis, the process of urban development generates concentrations of heat-absorbent materials that retain heat and create what are known as urban heat islands (UHI). UHI often have a higher temperature than their rural hinterlands. They produce lower air quality due to their production of ozone pollutants, and they may also provide lower water quality due to ecosystem stress as water temperatures increase as it flows into cities. It should be noted that UHI might not directly increase the overall mean temperature of the planet as a whole but concentrate some of the temperatures to certain urban spaces. Furthermore, not all cities have a significant heat effect, plus mitigation effects are being explored such as greening roofs to convert some carbon dioxide into oxygen and painting buildings white to reflect sunlight and absorb less heat.

Cities are situated more significantly as a place where consumption and production will generate significant levels of CO_2. Urban areas have even been described as a battleground for climate change (UN Habitat, 2011). Considerations for the contribution to global warming by urban areas are due to geographic location, demographics, urban form and density, economic activity and average income. Geographical location of urban areas is a key factor for climate change contribution, as it will determine the level of energy demand for heating, cooling and light-ing. To illuminate this example, a city such as Dubai, situated in a relatively extreme dry and arid environment, will have greater demands (in both transportation and generation) for heating in the evening, fresh-water consumption, use of coolants such as air conditioners, and lighting the rapidly expanding new-built environ-ment. These greater demands would be in comparison to a more temperate environment with existing energy infrastructure availability such as say Paris in northern Europe.

Location will not be the only contributor, as demographics such as the level of population of a city or town will contribute to the level of CO_2 as part of human contributions to global warming. Higher population density of a city, such as Manila in the Philippines compared to Detroit in the US, will, for instance, gener-ate a greater demand for space and services and contribute to CO_2 emissions. The urban form is therefore important as a more compact city rather than one with a considerable urban sprawl will be more efficient in providing a lower CO_2 average per capita. As well as consumption contributors, by consumers in cities, the produc-tion of goods and services within a city will be another factor adding to greenhouse gases. Concentrations of heavy industry are key contributors, whilst simultaneously generating high rates of economic activity. Chinese city examples of pollution and CO_2 contribution are abundant with the World Bank outlining 16 of the 20 most pollutant cities being Chinese, plus the city of Linfen, China, being voted the most polluted city on the planet (Blacksmith Institute, 2007) due to it affecting three million inhabitants because of its coal-powered energy and auto-mobile production.

The average income factor in urban areas is an interesting contribution to global warming. Analysis of this contributing income factor to consumption patterns can be applicable to both average and median income, especially as median incomes tend to be a common measurement in ranking city income. For a high-earning urban area such as San Francisco City and County, a median household income of $79,957 of its 798,000 residents (American Community Survey, 2006–07) would tend to create consumption patterns that generate high energy and resource demands. This would differ from a low-income base of an urban area such as Cleveland using US comparisons, where 250,000 residents in households earned a median income of just $26,535 (US Census Bureau, 2007). This would, in turn, generate lower energy and resource demands, and thus lower contributions to global warming.

The factors influencing the release of greenhouse gases from urban areas are, in reality, multifaceted and will have varying degrees of magnitude depending on the city example in question. For instance, using the US examples, per capita green-house gas emissions can vary between cities within the same country. For example, the UN Habitat report (2011) states that Washington DC, for example, has relatively high emissions proportionally as the city has a small population in relation to the number of office buildings for government and support functions. Whereas, by contrast, New York City's emissions are low, per capita, for a wealthy city in a developed country, owing in part to its high population density, small dwelling size, extensive public transport system and number of older buildings that rely on natural day lighting and ventilation. These factors, at varying degrees of magnitude, can also be compared between cities globally. From Table 3.1, it can be seen that Washington DC (US) holds the number one position at a global level in releasing 19.7 tonnes of CO_2 per capita in comparison to cities lower down the global list such as Beijing (China), which is sixth with a release of 6.9 tonnes of CO_2 per capita. The importance of analysis at city spatial scale, and hence the urban area, is demonstrated here as marked variances occur in greenhouse gas release by cities and their national boundary. Within the two global examples cited, the national release of CO_2 by Washington is slightly below the US national average of 23 tonnes of CO_2 per capita (2004–05), whereas Beijing is double the release with the national average of 3.4 million tonnes per capita (1994). National release per capita has subsequently risen in China as economic activity has increased exponentially during the 2000s, with reports claiming in 2006 that China (6,200 million tonnes) overtook (5,800 million tonnes) the US in CO_2 emissions (NEAA, 2008). Importance of demographics as one of the five contributing factors also prove important, as national population levels will be distributed disproportionately to the city in question – and thus the large national population of China outside of cities such as Beijing will produce far lower national figures at 3.4 CO_2 tonnes per capita.

The considerations of CO_2 release from urban areas are therefore relevant and will also connect to issues of climate change. Global warming from CO_2 release is often paired up with or seen as a significant cause of climate change. Climate change can be described as a clear, sustained change in climate over several decades or longer. The components of climate are those such as temperature, precipitation, atmospheric pressure or winds. Furthermore, such changes must constitute a clear trend and be clearly distinguished from the small random variation in these parameters that takes place all the time. The United National Framework Convention on Climate Change (UNFCCC, 2011) recognises multilaterally, over 192 countries at

Table 3.1 Comparing city and national per capita greenhouse gas emissions

City	GHG emissions per capita (tonnes of CO_2 eq.) (year of study in brackets)	National emissions per capita (tonnes of CO_2 eq.) (year of study in brackets)
Washington, DC (US)	19.7 (2005)	23.9 (2004)
Glasgow (UK)	8.4 (2004)	11.2 (2004)
Toronto (Canada)	8.2 (2001)	23.7 (2004)
Shanghai (China)	8.1 (1998)	3.4 (1994)
New York City (US)	7.1 (2005)	23.9 (2004)
Beijing (China)	6.9 (198)	3.4 (1994)
London (UK)	6.2 (2006)	11.2 (2004)
Tokyo (Japan)	4.8 (1998)	10.6 (2004)
Seoul (Republic of Korea)	3.8 (1998)	6.7 (1990)
Barcelona (Spain)	3.4 (1996)	10.0 (2004)

Source: Adapted from UN Habitat (2011)

the time of publication in which climate change and its effects need to be recognised. In defining the issue, 'climate change' is seen as attributed directly or indirectly to human activity that alters the composition of the global atmosphere and which is in addition to natural climate variability observed over comparable time periods (ibid.). Climate change can therefore be seen as the variability in climate, whereas global warming is seen as an increase in global temperatures measured over recent decades. To highlight such dual changes, it is stated by many governments that if it continues, climate change and global warming have the potential to seriously disrupt many of the environmental, economic and urban structures upon which human society depends (POA, 2008).

Contemporary evidence argues that recent increases in global temperatures in the late twentieth century and early twenty-first century are due to human release of greenhouse gases (GHG). Some results show that comparisons of climate model results with observations suggest that anthropogenic changes, particularly greenhouse gas (GHG) increases, are probably responsible for this climate change (Santer *et al.*, 1996). Even with the contribution to climate change from natural factors such as an increase in solar irradiance or a reduction in volcanism, the contribution to an increase in global temperatures and climate change are convincing. As an exemplar study, Crowley *et al.* (2000) argue that beyond natural influence is a clear case that human release of GHG is the most significant contributor to global warming. He states:

> I show that the agreement between model results and observations for the past 1000 years is sufficiently compelling to allow one to conclude that natural variability plays only a subsidiary role in the 20th-century warming and that the most parsimonious explanation for most of the warming is that it is due to the anthropogenic increase in GHG.
>
> (p. 279)

Studies such as these suggest that urban spaces with high human concentration are areas in which GHG are produced or consumed at a high level and worthy of explanation and understanding for some degree of mitigation or adaptation. Cities have therefore been at the heart of promoting adaptation and mitigation of climate change through various administrative institutions in urban localities. A few key alliances of local municipalities joining forces to lead on climate change include the ICLEI (Local Governments for Sustainability) (ICLEI, 2012) which is an association of more than 1,200 local governments from across the globe committed to sustainable development. There is also the Large Cities Leadership Council, also known as C40 Cities (C40 Cities, 2011). C40 was created in 2005 to forge alliances and co-operation among some of the largest cities from all regions of the world. Institutions involving key city agents include the World Mayors Council for Climate Change (World Mayors Council, 2011) – a 50-member international alliance formed in 2005 addressing climate, biodiversity and Millennium Development Goals (UN Habitat, 2011).

Due to the multiple agents and institutions involved in global warming and climate change, the local level has many complex urban issues that relate to the built environment and environmental resources. These local urban issues could be directly addressing global environmental concerns such as release of GHG or adaptation and mitigation of climate change at a household or neighbourhood scale. These local urban issues may also be cross-cutting with concerns for problems that are also pressing concerns for stakeholders in a locality – such as residents, community groups or local authority and city leaders. Cross-cutting urban issues include themes discussed in Chapter 11 such as: housing in developing affordable zero carbon homes; labour and employment in creating sustainable neighbourhoods in demand; education, health, crime and services in generating quality places with optimum well-being; transport and infrastructure in generating access via connectivity of urban spaces; and addressing inequality and poverty, which are essential for a just society. Initial thought in this area of study can be traced to ecological concepts of the neighbourhood (Park and Burgess, 1925) to more contemporary thinking on neighbourhood planning in the UK Localism Bill (CLG, 2011). To consider all spatial scales including the small-scale neighbourhood, any focus that involves the social, economic and environmental has intellectual association with sustainable development. This area of thought needs particular awareness in urban and environmental economics.

h. Natural resources and sustainable development

For urban areas to have future quality of life, their environmental influence will also have to interact sustainably with their social and economic drivers. Sustainable development is about making sure that people throughout the world can satisfy their basic needs now, while making sure that future generations can also look forward to the same quality of life. Sustainable development recognises that the three 'pillars' – economy, society and the environment – are interconnected (DEFRA, 2011). Sustainable development is therefore a noble aim but not seen widely in economic policy and practice, although lobbying and educational influence of the sustainable agenda has had much success. Defining what sustainable means is often a stumbling block for researchers and practitioners approaching this subject, although it may be more fruitful to consider the 'spirit' of the message – to maintain stability over a period of time. Confusions and clarity on sustainable concepts are recognised by Gilpin (2000) where he states:

> Sustainability, like apple pie and motherhood, is everyone's favourite. Business people are keen on sustainable profits, sustainable growth, and sustainable investments; academics may search for sustainable arguments; designers may seek sustainable performances. In the end, sustainability simply means keeping things going for a long period of time, perhaps indefinitely.

Sustainability is a slippery concept, as some phenomena are progressive for a moment in time and will only provide some momentary step change. The idea to sustain something is inextricably associated with time, so in a search to make things sustainable, place (such as urban space) may be pushed to the edges of the debate. Arguments can be brought out as to whether sustainability is inherently conservative or progressive in approach. On the one hand sustainability could be seen as conservative in that it has desires that keep the *status quo* and remain the same. This raises questions as to whether society wishes to sustain obvious socioeconomic inequalities – especially if the gap between rich and poor is sustainably widening. On the other hand, progressive arguments can also be brought into the sustainability debate, especially if a desire for sustainable communities such as Sustainable Communities Act, filtered into policy (CLG, 2007).

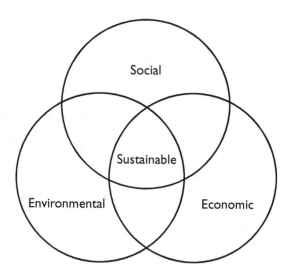

Figure 3.2 The three pillars of sustainable development
Source: Author

i. Pessimism and optimism: sides? The bright side and the dark side; current commentators

The environment according to economists, whether natural or built, is therefore well developed over several centuries and takes varying attitudinal outlooks. On the bright side, more positive and optimistic views tend to focus on the idea that urban and environmental improvements can be generated over time. As the future is ulti-mately uncertain, it is thought wrong to look at past events to predict an inevitable doom-laden future. The poor environmental record, the social problems faced by humanity and the widening social inequality may in fact be problematic (or even, fatalistically speaking, be inevitable). What is important is that solutions are thought through in either tackling the root cause or ameliorating its effects. For instance, the increasing demand for consumption in urban areas as population increases may be met by supply-side increases via technological advances or increases in economic growth. Further optimism can be brought via developments in recycling that may enable problems such as waste generated from urban areas to become part of a more sustainable future if managed in balance. Encouragement and incentives towards this brighter future can be presented both in ideas from educational spheres or in more direct action from the use of political pressure and the political will to change policy on environmental resources and issues, involving the built environment located in urban areas.

Economists would also use the argument of substitutes becoming a potential solution as certain energy sources used in urban areas become too expensive. A substitute is a good for which an increase (or fall) in demand for one leads to a fall (or increase) in demand for the other – e.g. substituting margarine for butter if butter becomes too expensive. A substitution effect in energy is said to occur if petrol has become cheaper relative to everything else, so people switch some of their consumption out of goods that are now relatively more expensive and buy more petrol instead. The reality in energy is that petrol and most other fossil fuels inflate in price over the long term, which means that cheaper alternative energy substitutes could become economically viable – hence the growth in renewable technologies such as windfarms. In essence, high prices in oil and gas will encour-age technological developments in renewables due to this substitution effect. This growth in renewable energy for consumption, largely in urbanising spaces, may be geographically varied. For instance, more developing nations may not have the capacity to research and develop their own green infrastructure and thus will incur higher substitution costs in making cleaner energy cheaper, and retain reliance on fossil fuels such as using charcoal fuels. Optimism can be taken forward, though, as trade and distribution of such resources can be multilaterally negotiated. Trajectories are not always set according to past events, especially as it can be seen that the Malthusian prophesies of restricted social improvement have not occurred at a global level, whilst there has been continued economic growth.

On a darker reflection, some economists can generate a more pessimistic outlook of urban development that more critically view that environmental resource use and social problems are being exacerbated. Often it is the ERE (environmental resource economics) approach that views environmental resources extracted and used in urban areas to be finite and lost for ever. The EE (ecological economics) approach is slightly more forgiving in thought as it allows the idea that natural resources have the ability to regenerate, be cultivated and be managed. However, once extraction of certain resources are passed tipping point, for instance in the over-cultivation of crops for food by residents in urban areas, the land may not have any future productive capacity. Therefore, in both approaches, some of the resource problems will not disappear unless some careful human involvement is made to reduce social conflict (in the case of social issues).

A bleak urban and environmental future is conceived if the Malthusian problems of population growth are taken to the forefront of the argument. In simple view, population growth aggravates resource competition and further enhances environmental crises. This aggravation may even be considered if there is a global economic slow-down, not just from further resource development in the boom time. It can be argued that such an economic stagnation or even recession may generate a switch to cheaper and less-environmentally-sound sources, and could further generate an unhealthy competition for resources that may endanger peace. More pessimistically, further environmental physical risks could result in tackling the energy question, especially if the substitution effect away from fossil fuels moves more towards nuclear power, and the higher associated risks of using this source. Switching of land use may also become a fiercely contested space, especially if energy needs from biofuel become more economically viable than agriculture. Interesting examples are that with the development of hybrid biofuel cars, this may make bread more expensive as the produce from agricultural land is used for fuel rather than food – what has been termed 'Burrito wars'.

Even more cynically with regards to the pessimistic outlook on urban and environmental economics, a raised social conscience towards attaining a sustainable development can be misappropriated for political gain. As an example, a city could make economic gains in the form of central government grants and private investment if it promotes itself as a 'green' city. But this may be to the detriment of the city's inhabitants if what is required is a greater focus on its social problems that may be more pressing. In this instance, local city officials may be playing the green game to appropriate wealth and power when their constituents are in need of a fairer distribution of wealth to enable a more sustainably developed urban area from a grassroots 'bottom-up' approach.

In weighing up both economic perspectives as both optimistic and pessimistic, the important creation of solutions may come from ideas beyond this duality. Claims for a constructive criticism may be more fruitful in a search for better quality of urban spaces that do not have detrimental impacts on the natural environment. Urban and environmental economics is not a question of sides. Lessons from ideas and thought in the past are important and just as relevant as new evolved ideas that have an eye on future cities and urban areas. What is needed and can be gained from this text is the need to research and allow careful study for a reasoned and rational balanced argument for change where necessary. Some changes will be sustainable; others will have less of a need to be. What is hoped by all progressive economic thinkers is that prosperity, whether as utility (in neoclassical economic language) or well-being (in more contemporary jargon), can be enabled for the greater good of people and the planet.

Summary

1   The origins of economic thinking can be conceived to date back to when humans bartered for goods or exchanged through a system of gifts.

2   Money as a commodity becomes part of the economic system, where, in particular, an 'IOU' or intermediary resource is needed to counter time delays and quantity restrictions of obtaining a good.

3   The system of commodity currency in many instances evolved into a system of representative currency such as gold and through the use of the gold standard.

4   Contemporary western systems of money are traded digitally as numbers on a computer screen. Bartering continues to be carried out within an estimated $8 billion-a-year industry.

5   Much of this thinking in the west has its origins in the Enlightenment (or age of reason) period from approximately 1650 to 1800, which promoted intellectual interchange and opposed intolerance and abuses in Church and state.

6   Towards the end of this Enlightenment period (eighteenth and nineteenth centuries) and entering a philosophical Romantic period (second half of eighteenth century), significant thinkers lived within adaptation to industrial development.

7   Romanticism was harking back to a period of nature as industry grew, population rose and urbanisation spread.

8   The Malthusian problem is whether per capita incomes would inevitably be driven down due to the rise in population levels. Critiques of this hypothesis is that (1) advances in technology may improve the productive capacity of food; (2) technological developments or education in birth control may also adjust the dynamics of population increase; and (3) wider attitudinal changes such as voluntary or regulatory restrictions of family size.

9   Ricardo's thinking was a conclusion that land rent (the reward for owning land) grows as population increases. This would generate, over a period of time, an aggregate appreciation in the value of property and necessity goods such as food.

10   David Ricardo was instrumental in developing the concept of comparative advantage. This theory has counter-weight problems in that the type of advantage chosen will form to meet the needs to the rich and powerful and not be a bilateral exchange. This is further complicated in that different resources for competitive advantage will be desired during different phases of the development process.

11   John Stuart Mill expressed that there were limits to such population growth globally, and he begins to bring in ideas of social welfare.

12   In welfare economics, Pigou (1877–1959) took Alfred Marshall's (1842–1924) *Principles of Economics* (Marshall, 1890) and the concept of externalities and embedding it further into the discipline. In order to deal with externalities, taxes and subsidies could provide incentives to negative and positive externalities.

13 Ronald Coase, in the 1960s, demonstrated that behaviours and actions outside of the invisible hand of the market and outside of the incentives of tax and subsidy could affect economic activity.

14 The Club of Rome report (1968), *The Limits to Growth*, recognises resource depletion, then subsequently highlights issues of global warming, climate change and sustainable development.

15 Concepts developed of significance include urban heat islands (UHI), demonstrating that cities are situated where consumption and production will generate significant levels of energy, heat and CO_2.

16 Significant considerations for the contribution to global warming by urban areas are due to geographic location, demographics, urban form and density, economic activity and average income.

17 Urban spaces with high human concentration are areas in which GHG are produced or consumed at a high level. Cities have therefore been at the heart of promoting adaptation and mitigation of climate change.

18 Sustainable development is about making sure that people throughout the world can satisfy their basic needs now, whilst making sure that future generations can also look forward to the same quality of life. Sustainable development recognises three interconnecting 'pillars' – economy, society and the environment.

19 As the future is ultimately uncertain, it is thought wrong to look at past events to predict an inevitable doom-laden future. For example, substitutes may develop as solutions as certain energy sources used in urban areas become too expensive.

20 A more pessimistic outlook on urban development more critically views that environmental resource use and social problems are being exacerbated.

21 Lessons from ideas and thought in the past are important and just as relevant as new ideas that have an eye on future cities and urban areas. Some changes will be sustainable, others will have less of a need to be if an insular outlook is to be avoided.

Chapter 4

The basic economic problem in shared spaces

Auckland, New Zealand

This chapter will now turn its attention to the underlying root cause or problem of shared spaces in order to further justify this subject area and try to tease out potential solutions to the ideas uncovered by past and present economic thought. As a discipline, economics attempts to question fundamental issues that could be applied to several spaces of inquiry. The basic economic questions centre mainly on three basic problems of daily living, which every individual or social group must attempt to answer. In short these questions are:

1   *What* goods and services to produce?
2   *How* to produce these goods and services?
3   *For Whom* to produce these goods and services?

In an urban area, '*what* to' produce would include goods such as housing, and services such as shops and healthcare. With the example of housing, it has the characteristic of being a consumption good, a social good and an investment good. Therefore, *what* housing is produced, is on the basis of a combination of its characteristics generating from varying forms of tenure, whether private rented, social rented (from a local authority or housing association) or owner-occupied.

'How to' produce goods and services in urban areas will be determined largely by some combination of market signals and government intervention. The forces and decisions on what goods and services will be produced can therefore vary depending on space, with different cities being subject to different contextual circumstances. For instance, using a simple urban housing example, a city with a greater national free-market ethos will tend to construct more dwellings as the market economy rises (and *vice versa*). A city located in a nation with a greater state-controlled and less of a free-market approach would construct housing in more of a regulated and planned way. For example, how housing growth is produced could be by allowing units of certain housing to match growing employment growth in certain cities and regions.

With regards 'for *whom*' these goods and services are realised, and by which method, different social groupings may benefit or lose depending on particular social cleavages such as wealth and class (lower, middle, upper, under), race, religion, gender, age and sexuality. Particular minorities and majorities will benefit from the choices made in society – including private decisions by, for instance, commercial property interests, and public interests such as politicians seeking the best combination of public returns for their constituents. Different goods (e.g. the production of merit goods such as education) and how they are produced (e.g. public, private or a combination) will affect who ultimately receives the good or service. As an example, not all schoolchildren will be able to afford private education.

So, more broadly, economics is the study of how society decides what, how and for whom to produce (Begg *et al.*, 1994). It is the decisions made by society that are of importance in determining the three questions of what, how and for whom to produce – and these decisions can often be influenced by the level of wants and resources available.

a. Resources, wants and scarcity

At either end of the spectrum, in which wants and resources are situated, is the idea of 'infinite wants' at one end, and 'scarce resources' at the other. Infinite wants holds the notion that people are hard-wired to always want to consume more goods, services and experiences. Scarce resources in the form of factors of production (land, labour, capital and entrepreneurship) restrict the ability for all of these infinite wants to be satisfied completely – as scarcity defines the resources as having limited availability. Due to this mismatch of infinite wants and scarce resources, it is thought, in economic terms, that a choice will need to be made. Using an urban and environmental example, a situation can be described where residents will want to live in property that allows improved quality of life – such as whether that is a better location, better school catchment, areas with low crime, commutable to suitable employment and near to recreation space. However, these wants are curtailed due to the scarcity of resources. For instance, there is only finite (or even unique) land available for desirable locations, there are a limited number of available schools, not all locations are crime- (or fear of crime) free, the commute time to work will have a limit, and there is only a finite amount of land available for recreation. As a result, this housing example demonstrates that the wants will not be completely satisfied given the scarce resources (and factors of production) and a compromise and choice will have to be made. This idea of choice can be aggregated to a national scale and demonstrates that not all of a nation's inhabitants will be satisfied, a thought that is held in that:

> The production that can be obtained by fully utilising all of a nation's resources is insufficient to satisfy all the wants of the nation's inhabitants; because resources are scarce, it is necessary to choose among the alternative uses to which they could be put.
>
> (Lipsey and Chrystal, 1995)

Mention has been made of scarcity resources and its relation to factors of production. These will need to be unpacked to provide more precise economic analysis of urban areas and their connection with environmental resource issues. The types of resources that are under scrutiny as connected to urban areas involve resource categories that are (1) natural, (2) human or (3) capital. Natural resources would be raw materials supplied by nature such as forests, fresh water and rocks. Human resources would be the people who work to provide goods and services such as construction engineers. Capital resources would be the tools, equipment and buildings that are used to produce goods and services for future production, such as the cranes and lorries used to develop residential property sites.

A separate categorisation may be more analytical for economic purposes of producing goods and services. This categorisation is known as the factors of production in which there are four. They are (1) land, (2) labour, (3) capital and (4) entrepreneurship. The four factors of production are a requirement to produce a good or service. Land includes natural resources, as in mining, and is an increasingly important factor in urban areas as more generally land for (re)development is more scarce than in rural areas. Labour includes all human resources. It may be unskilled, semi-skilled or skilled, and local labour markets vary in the size and nature of the pool of labour in urban areas. Cheap, unskilled and semi-skilled labour may be an important locational factor for multinational corporations while skilled labour is significant in high-technology industries. Capital covers all manmade aids to future production; fixed capital includes the physical plant, buildings, tools and machinery; while circulating (or working) capital includes raw materials and components. Entrepreneurship as a factor of production is closely aligned to labour but often takes on its own unique factor, as it takes special human qualities that can manage and combine the other three factors of production effectively, particularly in taking on risk in new products and methods of production.

The element of risk is important and is a particular input by an entrepreneur that will be expecting some element of reward – often in the form of profits. Each factor of production in this way has inputs such as risk-taking, with rewards such as profits. Similarly, with the input of land as a factor of production, the output reward could be rents. For labour, the output reward could be in the form of a salary or wage. Capital inputs as a factor of production could reward in the form of financial interest returns.

It should be held that this combination of the factors of production to produce an end-product as good or service considers only the supply-side considerations and will only maximise reward if the product is to meet wants for consumption. In essence, the production of goods and services is only as a 'means to an end', to meet the wants of people, and there will only be an incentive to produce if they will be consumed. In a perfectly competitive market the consumer is seen to be 'king' rather than producer – this will be covered further in later chapters.

b. Choice and opportunity cost

Choice is therefore an important element in economic thinking if there are scarce resources and infinite wants. Urban areas that are booming will have to make choices, as the resources needed to sustain development will be in high demand but not freely in supply. In residential development, the use of materials may have been formed by an economic choice based on the cost – such as due to the high price of steel in production methods. In essence, choice is generated because scarce resources cannot meet infinite wants. Classic economic texts would say that choices are necessary because there are insufficient resources to satisfy all human wants (Lipsey and Chrystal, 1995). This is also why the discipline of economics is, on occasion, refereed to as the 'science of choices'.

In exploring the key economic questions further, the question of 'for whom to produce goods and services' is important. There will be a choice made by society as to whose wants will be satisfied and whose will be left unsatisfied. This is because different people and organisations at different periods of time make choices. These choices could be made both individually and collectively by different stakeholders such as consumers, businesses, unions and governments. For example, governments could decide to control and incentivise the level of development on Brownfield sites by investing in regeneration for those areas previously experiencing industrial decline. This, in turn, would aid less prosperous and declining regions and provide the potential to satisfy some of the wants of disadvantaged places and people that would have not been satisfied without a redistribution of resources.

The level of influence in choice by these people and organisations differs among economies. In the regeneration example, market intervention to act as a catalyst for a redistribution of resources will only be possible if a nation's government has the political will and resources available to enable such a course of action. This raises questions as to whether there is always a choice in society. For a particular city, there may not appear to be a free choice with respect to a household choice of school, healthcare, employment, housing, safety and local environmental quality. Some disadvantaged households could claim that they incur multiple hindrances, possibly so much so that economic and community constraints inhibit the choice to move to a more prosperous location. Ultimately, it can be argued that there is a choice made by society to enable those without a choice to have one. The availability of one vote per person is a solid foundation on which lies free democratic choice in western political systems.

Economically speaking, rather than politically, opportunity cost is one concept that can be used to demonstrate how economic choices are made. Opportunity cost is seen as the cost of forgone alternatives. As an example, the opportunity cost of public money being put into a public road is the cost forgone in providing more public funding to schools. In short, more roads will mean a choice to have less schools, a £500 million motorway extension has an opportunity cost of a £500 million publically funded school. The opportunity cost is the £500 million school cost forgone. Choices therefore involve alternative courses of action that are decided upon, and these alternatives may not simply be one or the other but alternatives that could be ranked in best preference. A more precise measurement of opportunity cost could therefore be the sacrifice of the best alternative choice. If the next best alternative to building a road is to maintain the land as wetland, the opportunity cost of building the road is the sacrifice of the wetland. To place some sort of value on the opportunity cost of sacrificing a wetland, some economic tools will need to be used, such as cost-benefit analysis, as demonstrated in subsequent chapters.

So the questions of choice with regards to what, how and to whom goods and services are produced are attributed to the fundamental basic economic problems. However, these major questions take differing magnitudes and focus depending on the economic scale of enquiry – microeconomic or macroeconomic scale. Theory of microeconomics and macroeconomics and their application to urban and environmental issues will now be unpacked to demonstrate the varying nature of enquiry that can be covered. As definition, microeconomics engages with the branch of economics that studies the economy of consumers, households or individual firms. This is different to macroeconomics, which, by definition, is the branch of economics concerned with aggregates, such as national income, consumption and investment. For urban and environmental economics, microeconomics would focus on, say, disaggregate household consumption of fuel within a particular city, whereas macroeconomics would focus on the aggregate national output of completed construction projects.

Microeconomics concentrates at a more disaggregate scale and can therefore provide specific detail on products, the methods used, distribution and efficiency. The specific products can be analysed with regard to what and how much of particular items are being produced. This could be how many cars are manufactured at a specific plant in a city. The methods used could entail the way in which the assembly line is organised at such a manufacturing plant, which may affect the flexibility of employment practices. Distribution of goods and services can be analysed under microeconomics, and especially as to how they are divided as a proportion to different strata of society – such as analysis of who has access to the consumption of car use. Within regards to efficiency, this type of microeconomic analysis can begin to provide a better understanding of whether these production decisions, which have been made, are giving good outputs considering the inputs into the process. Using the city car-manufacturing example, do the factors of production going into the process (land site, skilled and unskilled labourers, capital – machinery, entrepreneurship – CEO) provide a greater value in outputs such as added site value, a living wage, capital returns or company profits.

Macroeconomics takes a bigger, broad-brush picture by aggregating many microeconomic concepts and concentrating on wider market forces. Three key macroeconomic concepts involve utilisation of resources, the influence of inflation and importance of capacity. With regards to utilisation, the issue centred on is whether resources are actually being used. There could be a situation where labour, for instance, is being under-utilised and, as part of the wider macroeconomic picture, could be having detrimental economic consequences for society. Unemployment and lay-offs at a car plant in a major industrial area may make microeconomic sense in terms of efficiency but could be a significant macroeconomic cost in terms of wasted labour, benefit costs and costs to socioeconomic well-being for potential future productivity.

Inflation as a macroeconomic concept looks at what are the causes and consequences of the change in the purchasing power of money. Inflation, more specifically, is the overall general upward price movement of goods and services in an economy, usually measured in the UK by the Consumer Price Index and the Retail Price Index (in the US as the Consumer Price Index and the Producer Price Index). If there is an upward movement in the price of goods and services it will mean that the currency used (e.g. pound or dollar) would have less value because fewer items in the same standard basket of household goods would be able to be purchased. To use an example from the built environment, if the price of houses began to increase, it would mean (*ceterus paribus* – other things remaining equal) that fewer houses could be bought for the same amount of money – in effect, relative to house prices, money has lost some of its value or purchasing power. For those that previously owned property, they will have gained some of the rewards from house-price increase, which will balance out against any relative losses in the purchasing power of money.

Capacity is the third key theme of concern for macroeconomics and takes into consideration a wider view – what is the economic capacity for growth of goods and services. The economic output of some cities will vary in terms of how much product or service can be generated given the factors of production available. For instance, there is a current capacity for a city to manufacture goods such as bicycles and computers, or provide services such as housing maintenance. Taken at an aggregate, macroeconomic scale, the capacity for a city to enable skilled service production, say in terms of web development, will depend on the number of available skilled workers available in that city. Macroeconomic considerations, therefore, can look at themes such as utilisation, inflation and capacity; at wider geographical scales higher than the firm or household but also lower than national boundary scales such as at the city scale.

Output at the city scale is often measured in terms of GVA (gross value added), which is an aggregate productivity metric that measures the difference between output and intermediate consumption. GVA provides a currency value for the amount of goods and services that have been produced, less the cost of all inputs and raw materials that are directly attributable to that production. In essence, GVA can reveal, at a city scale (but also at a microeconomic scale for an individual firm), how much money the products or services contributed towards meeting companies' fixed costs and providing opportunity for a bottom-line profit.

As in the example of GVA for urban areas, measures of value can be of concern to both microeconomics and macroeconomics. Alternative economic scales of enquiry can be used within the discipline, especially as we have demonstrated that some economic phenomena can affect both microeconomic concerns, such as the effect of inflation on a good or service for a firm, as well as the impact of inflation on a wider national economy. A further economic scale of enquiry is what has been termed the 'meso-economic' scale of enquiry. Meso-economics is a term that is used to describe the study of economic arrangements that are not based either on the microeconomics of buying and selling and supply and demand, nor on the macroeconomic reasoning of aggregate totals of demand. It therefore takes more of a structured approach, but one that is possible to measure, and dates from the 1980s where it was questioned whether there would ever be a bridge between the two main economic paradigms in mainstream economics. To use a meso-economic example from urban and environmental issues, it would involve analysis of unemployment in an urban area. Here, for example, information asymmetry, where one party has more or better-quality information than the other party, is used to gain better employment prospects, contrary to normal economic reasoning where actors gain by increasing the rate of dissemination of information. Information and knowledge as wealth and power is discussed in terms of behaviours between the different parties, rather than microeconomic analysis of supply and demand of information, or what the aggregate totals of this supply and demand are for a bounded subject of enquiry such as a city or a nation. The co-operative and competitive (or even evolutionary) nature of a city's actors and organisation, determining what natural resources will be utilised, could therefore be one area of meso-economic urban and environmental enquiry.

c. The production possibility boundary

In moving back to classical microeconomic forms of enquiry in urban and environmental economics, the use of production possibility boundaries (PPB) enable a simple rational explanation of what, how and to whom resources are produced in society. Furthermore, as the title suggests, such PPB analysis can show what the limits of production are and what choices are possible. A classic textbook example, which is replicated for this text, is a demonstration of the production possibility for two goods or services that could be produced given the available factors of production, that of either military goods (e.g. guns) or civil goods (e.g. food). As per usual with economics, a simple model as a thought experiment is adopted, and thus a notion of *ceterus paribus* is held (other things remain the same). This is to aid in deconstruction of economic elements in society to enable better understanding of what the constituent parts are and how they interact with each other under different circumstances. Philosophical flaws to this approach are not within the scope of this book, but it is assured that this line of enquiry helps in building basic economic thinking and how it can provide improved analysis and discussion of urban and environmental matters.

To explain the basic choices and limits for production in society, Figure 4.1 diagrammatically demonstrates some of the economic forces previously discussed in action. The diagram explains in a duality the relationship between choosing a greater quantity of civil goods, as demonstrated in the vertical axis, or a greater quantity of military goods, as demonstrated in the horizontal axis. The production possibility boundary (PPB) is represented by the arc in the diagram, and represents the maximum quantity of both goods that can be produced in an economy. If there is a quantity of military goods at point M0, it will enable a quantity of C0 civil goods to be produced, with the production possibility of both goods being produced at point 'a' (high civil goods and low military goods). Consequently, if there was going to be a greater quantity of military goods to be produced to M1 (e.g. such as by a change in government contracts), this would enable a lower quantity of civil goods to be produced at point C1. The production possibility for both quantities at this new balance would be at point 'b' (low civil goods and high military goods). The PPB does not always represent what is actually being produced in an economy; it is only the maximum possible amount of goods and services able to be produced at the most efficient process at a fixed point in time. The actual production of both military and civil goods could be at, say, point 'c' (attainable inefficient combination) where there is less of both goods and services produced, but not all of the input factors of production are being used at their optimal rate to produce the maximum output of goods and services for both sectors. Conversely, it could be conceived that there is a situation where a greater quantity of both military and civil goods is produced than is possible at point 'd' (unattainable combination). This is indeed unobtainable as any combination of factors of production would not be able to produce these quantities, that is unless the whole production possibility boundary can shift outwards (see Figure 4.1).

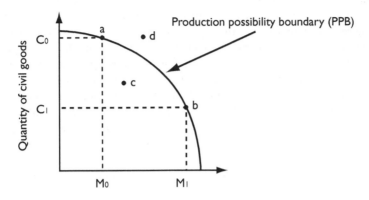

Figure 4.1 Civil vs. military goods
Source: Author

The PPB curve plots maximum attainable combination of all resources:

d = Unattainable combination
c = Attainable inefficient combination
a = High civil goods and low military goods
b = High military goods and low civil goods.

It should be picked up here that urban and environmental issues could be easily attached to this economic concept and model. The combination of, say, residential or industrial property in a city could be analysed rather than the production of military and civil goods. As space is finite, only so much land can be used as a factor of production, unless property is built upwards, and again, this is limited to some degree. In a simple dual model of urban land use as residential and industrial, there will be different possible combinations of production for each sector. There could be more residential property to the detriment of industrial property (point a or b on the curve); both sectors could produce less if other factors of production beyond land (e.g. labour, capital, entrepreneurship) are not utilised efficiently; plus there is a possibility beyond the curve for increases in both sectors of property if, over time, the economy for an urban area increases in growth.

In developing PPB models further, the ability for boundary curves to shrink and expand over time begins to allow us to understand how factors of production can be expanded as economic growth expands or contracts. This economic growth could be, for instance, the growth in GDP (gross domestic product) for a nation where it is statistically evidenced that output of goods and services in a country are expanding. For urban areas such as towns and cities, the increase in economic growth could be measured in GVA (gross value added) and would indicate that either more factors in a fixed urban space are being utilised or the urban area is being expanded in space as the built environment develops further. Figure 4.2 diagrammatically shows, using a PPB, how an economy's capacity to produce goods and services can expand (note it can contract) over time to enable more production possibilities of the two groups of goods in question. The solid arc has production possibilities 'a' and 'b' at a first period of time (say year 1) for either civil or military goods. Then, with an expansion in the economy to a second point in time (e.g. say year 2 – through increase in GDP or GVA), there is a new production possibility boundary that has shifted outwards at point 'd', which is now attainable after growth. In essence, as a result of this growth, it can be seen that more of all commodities can be produced.

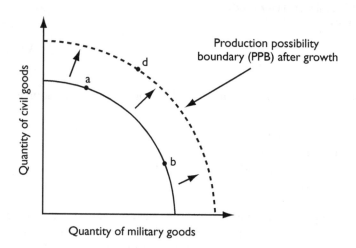

Figure 4.2 Production possibility boundary (PPB): Outward movement following economic growth

Source: Author

What Figure 4.2 also demonstrates is the use of some of the economic concepts previously introduced being used in action. The central concept of scarcity is certainly apparent in the PPB model as the boundary arc represents the scarce resources available to urban areas at a particular point in time. In Figure 4.2, point 'd' is unattainable at the first time period (the solid arc) and therefore the limits of scarce resources exist at the boundary. Economic concepts of choice are also demonstrated in Figure 4.2, as on, say, the first point in time (the solid arc), given a set amount of economic capacity, a choice can be made in society whether a combination of goods at point 'a' or point 'b' is desirable. Opportunity cost is another economic concept discussed that can be applied to the PPB model – the cost of the next best alternative forgone. In Figure 4.2, the concept of opportunity cost can be represented as the negative slope of the curve (or arc). For instance, if there begins to be an increase in military goods, the opportunity cost of the next best alternative (for which there is only civil goods) is the cost of the civil goods that are going to be forgone.

Further macroeconomic concepts can be discussed with relation to the PPB diagram as set out in Figure 4.3. With regards to utilisation of resources, at point 'c' it could be argued that resources are being under-utilised. This can be, say, the factor of production of labour that is unemployed for a particular city. In order to get the unemployed resources (such as labour) employed and utilised, it will take a movement from point 'c' to 'a' that may have been enabled through inefficient resources becoming more efficiently used. Using the labour example, this could be the supply of labour becoming more skilled and therefore more efficient in more skilled tasks – assuming the demand for employment in an urban area is for skilled labour. As previously mentioned, the movement of resources from point 'a' to point 'd' is related to the ability for an entity such as a nation, or in an urban and environmental case, the space of a city or town to increase economic growth and therefore its capacity to produce more of all goods and services.

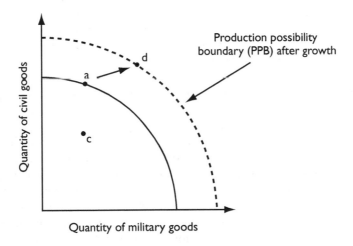

Figure 4.3 Production possibility boundary (PPB): How is it possible to get more of everything?

Source: Author

The use of PPB models can also be extended from the use of specific goods and services to their production within the public and private sectors. The separation of goods and services into two distinct spheres is difficult as many goods and services will have an overlapping economic and financial structure. For instance, a transport rail network may be owned by the public sector but the trains that are licensed on the track may be owned be privately owned competing companies. A separation can be made for modelling purposes. It is held that, from an accounting point of view, all goods and services will have some form of ownership and valuation by both public sector and private sector organisations.

Figure 4.4 demonstrates the PPB of private and public (or social) goods that can be provided by a particular bounded entity such as a nation or urban area. In this simple example a choice is made (through various free market and planned forces) in terms of the production of private goods, such as private cars and computers, or in the production of public goods, such as a public health service, public police force, social housing or state education. The extent of the PPB will provide a conceptual limit to what optimum amount of public and private goods can be produced in some sort of combination. This combination could involve a greater amount of public goods in relation to less private goods (e.g. at point 'a') or a greater production of private goods in relation to less public goods (e.g. at point 'b').

In applying real national economy cases to this public–private dichotomy, a country with a greater proportional production of public goods would be, say, Sweden (at point 'b'), and a country with a greater proportional production of private rather than public goods would be, say, Japan (at point 'd'). More mixed economies, such as, say, the United Kingdom, with a more balanced mix of public and private goods and services such as public (National Health Service) and private healthcare options, would be representative of point 'c' on the diagram. As greater private free market forces are introduced to a country the dynamics of the model will tend to show a shift of position along the PPB curve. Economic growth (measured in GDP) will be seen as previously with a shift of the PPB curve outwards over a period of time – hence more public and private goods and services are seen to be produced.

The choices of combination in the amount of public and private goods are not necessarily an individual economic choice. Decisions by society on the correct balance of goods and services are often determined via the political process. For instance, the decision on whether healthcare is public or private will, in part, be determined on the strength of private market interests, public support for certain sector control and the policies set by elected public officials. In addition to public and private concerns, overall growth by both sectors will have implications for what resources are used in the production process and determine to what degree these finite resources are utilised. With regards to environmental resources, the returning question of scarcity is shown, as economic growth represented by the PPB will have a limit to how far it can expand. It may be argued that there is a limit to economic growth due to the reproductive capacity of environmental resources being constrained, and especially finite at a global scale. The limit(s) to growth is the focus of attention in the next chapter.

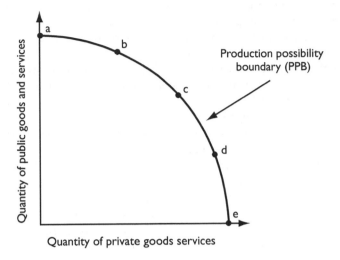

Figure 4.4 Production possibility boundary (PPB): Public and private goods and services
Source: Author

Summary

1   It is the decisions made by society that are of importance in determining the
    three questions of what, how and for whom to produce – and these decisions
    can often be influenced by the level of 'wants' and 'resources' available in rela-
    tion to urban space.
2   Choice is therefore an important element to economic thinking if there are
    scarce resources and infinite wants. Different people and organisations at differ-
    ent periods of time make choices. These choices could be made both
    individually and collectively by different stakeholders such as consumers, busi-
    nesses, unions and governments. It can be argued that there is a choice by
    society to enable those without a choice to have one.
3   Opportunity cost is one concept that can be used to demonstrate how
    economic choices are made. Opportunity cost is seen as the cost of the next
    best alternative forgone. As an example, the opportunity cost of public money
    being put into a public road is the cost forgone in providing more public fund-
    ing into schools.
4   Shared urban space and resource allocation questions take differing importance
    and focus depending on the economic scale of enquiry – at the micro-
    economic or macroeconomic scale.
5   Microeconomics concentrates at a more disaggregate scale and can therefore
    provide specific detail on products, the methods used, distribution and effi-
    ciency.

6   Macroeconomics takes a bigger broad-brush picture by aggregating many microeconomic concepts and concentrating on wider economic forces. Three key macroeconomic concepts involve utilisation of resources, the influence of inflation and the importance of capacity.

7   Note that measures of value can be of concern to both microeconomics and macroeconomics, such as in the example of GVA for urban areas.

8   Meso-economics takes more of a structured approach, such as exploring the co-operative and competitive (or even evolutionary) nature of a city's actors and organisation determining what natural resources will be utilised.

9   The use of production possibility boundaries (PPB) enable a simple rational explanation of what, how and to whom resources are produced in society. Furthermore, as the title suggests, such PPB analysis can show what the limits of production are and what choices are possible.

10   For urban areas, the increase in economic growth could be measured in GVA (gross value added), and would indicate (by PPB boundary expansion) that either more factors in a fixed urban space are being utilised, or the urban area is being expanded in space as the built environment develops further.

11   The use of PPB models can also be extended from the use of specific goods and services to their production within the public and private sectors. The separation of goods and services into two distinct spheres is difficult, as many goods and services will have an overlapping economic and financial structure.

Chapter 5

Limits to growth

Balancing space and resources

Budapest, Hungary

Limits to growth have underpinned initial considerations for urban and environmental study. In order to analyse these limits, a determination of what growth is will first be explained. From this, the restraining forces at work on growth are then revealed and placed within a framework, prior to discussing what limits can more broadly be considered. As a thought experiment, a notion of what growth could be envisaged without constraints will bring out further parameters to environmental resource allocation for urban areas. Further issues on the limits to growth are concluded, with a particular focus on developmental imbalances.

a. What is growth?

Growth, or more precisely economic growth, is an important area of focus when considering urban and environmental issues. Economic growth in urban areas will no doubt have consequences in the amount of resources that is being processed. Some of these resources will be extracted from the 'natural' environment and thus transform the built environment. With reference to Kohler's (1973) view of mankind on a crowded spaceship called Earth, the most important long-run goal is in maintaining a high-quality spaceship that has high-quality physical and human capital, and doing so with a minimum of investment and consumption. Urban areas are certainly an area to watch in order to maintain quality, and as stated (ibid.), maximum real growth will only be a true source of pride if it is not brought at the cost of environmental destruction.

The measure of economic growth is often set within a particular boundary at the national or 'domestic' spatial scale. A variety of country or regional measures of national income and output to determine whether growth is occurring are used in economics, including gross domestic product (GDP), gross national product (GNP), gross national income (GNI), and net national income (NNI). Gross domestic product (GDP) is the most common national measure and is concerned with statistically measuring everything produced by all the people and all the companies in a particular country. In the US, the Bureau of Economic Analysis (BEA) records and disseminates the production value over a specific period of time. As an example, in 2010 the US, GDP figure for the year was $14.7 trillion (in long hand that is $14,700,000,000,000). In the UK, the Office for National Statistics (ONS) records the official statistics for GDP, and as a comparator, the UK in 2010 had a yearly GDP figure of $2.3 trillion ($2,300,000,000,000).

In the UK, three different theoretical approaches are used in the estimation of one GDP estimate. GDP (O) from the output or production approach measures the sum of the 'value added' created through the production of goods and services within the economy (i.e. production or output as an economy). This approach provides the first estimate of GDP and can be used to show how much different industries (for example, agriculture) contribute within the economy. GDP from the income approach – GDP (I) measures the total income generated by the production of goods and services within the economy. The figures provided break down this income into, for example, income earned by companies (corporations), employees and the self-employed. GDP from the expenditure approach – GDP (E) measures the total expenditure on all finished goods and services produced within the economy (ONS, 2012). So, in short, there are three ways of calculating GDP – all of which should sum to the same amount since the following identity must hold true:

National output = national expenditure (aggregate demand) = national income

It is also important to note for the UK, that GDP includes the output of foreign-owned businesses that are located in the UK following foreign direct investment in the UK economy. The output of motor vehicles produced at the Nissan car plant on Tyne and Wear and the output of the many foreign-owned restaurants and banks all contribute to a country's GDP. In further analysing and breaking down GDP, the estimates are 'gross' because the value of the capital assets actually worn away (the 'capital consumption') during the productive process has not been subtracted. For example, the depreciation costs of the plant and machinery at a car manufacturers have not been taken into consideration, meaning the value added is the value of the potential sales of cars that are put onto the market for consumption. Secondly, the 'domestic', as stated, refers to the national or country-specific boundary being measured. Thirdly, the 'product' is in terms of the value-added output of goods and services (GDP (O)); or the income received to companies, employees or the self-employed (GDP (I)); or the expenditure as an aggregate on all products and services that have been produced (GDP (E)).

Taking into consideration the value of the national economic production based on some ownership of resources, it generates gross national product (GNP) as different to GDP. GNP measures the output produced by enterprises such as those owned by nationals in the form of labour or property, rather than GDP, which measures output produced within a particular geographic boundary. So in the case of foreign direct investment (FDI), if a particular firm's resources (e.g. property) are owned outside of the given territory, the GDP figure for the domestic nation geographically bounding the firm would be less than the GNP figures, as some of the resources will not be measured under GNP for the domestic nation being accounted for. GNP is therefore a value that includes elements of GDP that are owned nationally, plus any income earned by domestic residents from overseas investments, minus income earned within the domestic economy by overseas residents.

The final growth measure in addition to GDP and GNP is net national income (NNI) – that is gross national income (GNI) less taxes. GNI is the production for consumption of all goods and services over one year with the addition of gross private investment, government consumption expenditures, income from assets abroad and net exports. GNI, therefore, more generally includes the production of all products and services, plus the income generated from overseas mainly in interest and dividends. With a net income consideration, NNI takes into consideration the GNP (i.e. production owned by nationals – rather than within the geography of a nation) but deducts the business taxes paid to the government of the producer nation. Hence, this generates a net income value via an expenditure method of national income accounting. NNI more technically encompasses the income of households, businesses and the government. It can be expressed in a formula as:

NNI = C (consumption) + I (investments) + G (government spending) + NX (net exports: exports minus imports) + (net foreign factor income) – (indirect taxes e.g. corporation tax) – (depreciation)

GDP (and GNP and GNI) are often more specifically recorded statistically as 'per capita' GDP (and GNP and GNI). This means the total value of production at a domestic (by geography) or national (by nationals), or the income by nationals, is divided by all of the population for a particular country. When economic output is divided by the average population, the measure is recorded statistically and referred to as purchasing power parity (or PPP). PPP as estimates generated by the World Bank (WB) and the International Monetary Fund (IMF) are arguably more relevant in comparing national wealth. This is particularly so, as the PPP figure also takes into consideration the relative changes in inflation rates and the costs of living, rather than just simple exchange rates that may distort the real rates of income. In the application of using per capita economic output or income values (measured by PPP), it would be found that if an economy was growing but the population was growing at a faster rate, the average incomes for that particular country would be falling as there is less wealth to share among its nationals.

Considering economic growth in terms of rates of change is important if analysis of the dynamics of growth and its limitations are to be realised. If GDP is statistically providing a value at a point in time such as $14 trillion in 2010, this value may increase or decrease over the course of another time period, such as 2011, to say $15 trillion for the whole year. As a percentage change, or the rate of change as a percentage, it is calculated as:

{New value (2011 = $15 Tr)] − [original value (2010 = $14 Tr)/ original value (2010 = $14 Tr)]} × 100

= [(15 − 14) / 14] × 100 = 7.14%

This simple calculation would explain that there was a 7.14 percentage point increase in economic growth over a year for the country in question. In using GDP as the key measure of economic growth, 'growth in the economy' is often provided by quoting the growth in GDP during the latest quarter. So, an annual rate is calculated here in terms of the total output change in value compared to the previous quarter, expressed as a percentage change. A specific year for the same recorded example would be the gross domestic product (GDP) in the United States, which has expanded 1.9 per cent in the first quarter of 2011 since the first quarter of 2010. To get some sense of wider rates of growth, from 1947 to 2010, the United States' average quarterly GDP growth was 3.30 per cent, reaching a historical high of 17.20 per cent in March 1950 and a record low of −10.40 per cent in March 1958.

Negative rates of GDP provide a technical presentation of an economy being in recession. A period where there is neither growth nor a fall in output (or in some cases at a constant low of less than 2 per cent) is where an economy is described as being in stagnation. Prolonged periods of stagnation or recession would tend to generate actions by governments to boost output. These actions may differ and will have disproportionate consequences for different urban areas, as the fall in economic output will be different in different cities whilst a national GDP economic growth indicator will be showing a uniform negative growth rate. It is therefore important for urban and environmental economics to scale down analysis and discussion of different urban growth rates as well as scaling up to consider what environmental resource issues in relation to growth can be understood and tackled at a more international and global level.

b. Key restraints on growth

The key restraints on growth will have to be explored at differing scales of economic growth as different restraints will have different magnitudes of control. For instance, at a household level a significant restraint on economic growth would be the number of households in an urban area that are highly skilled and therefore generate higher household incomes and added output value. At a global scale of economic growth, which is in essence the economic growth of the entire planet, it will be restrained by the ability (social, political, economic, physical) to extract natural resources. Natural resources are one of seven categories to consider as the key restraints on growth that cover a multitude of geographic scales as shown in Table 5.1.

Table 5.1 Key restraints on growth

Restraint	Explicators
1. Natural resources	Topography Fertility of land
2. Agriculture	Size of agricultural base Subsidies
3. Population growth	Consumption Skills
4. Culture	Consciousness and awareness Education
5. Access to finance and capital	Savings Investment funds Credit
6. Debt	Domestic and international Repayments
7. Infrastructure	Facilities Services Installations

Source: Author

In discussing the key restraints on economic growth, natural resources are firstly seen to be a key constraint as it is a finite physical structure, given some room for regenerative capacity of some environmental resources (such as the regenerative ability of forests to produce more materials such as wood and rubber). As well as the fact that natural resources, being finite, stifle economic growth, the characteristics of natural resources will, too, constrain growth. For instance, the topography of cities that generate economic output through residential property development could be inhibited if it is extremely mountainous and cannot be built on due to problems of laying substantial foundations. In order to serve the consumption needs of urban areas, the fertility of natural resources will also be a constraint on how many materials can be produced and be an input into the productive process to develop the built environment.

Secondly, constraints to economic growth by agriculture will be important as the production and consumption of food as well as other materials (such as for the built environment) will need to be considered in relation to urban areas. At a global scale, if the agricultural base is not able to feed the current global population, then it could be assumed that there is less potential for economic growth due to human resources not being able to flourish efficiently, e.g. by being malnourished. The disproportionate distribution of agricultural produce is realised, so the restraints on growth will vary if comparing different urban areas that consume edible produce. Imbalances that generate further constraint with regards to agriculture could be the level of subsidy that is allocated by governments to enable more production of certain food goods such as through subsidy on fertiliser or on particular food products such as wheat. At a more general level, more economic incentives for agriculture will tend to put less of a constraint on economic growth for the domestic nation and its urban areas.

Population growth is the third key constraint to growth as it can be argued that a space with high population growth could generate more constraints. This would depend on whether economic growth is largely being generated by labour-intensive industries; in the case of mostly manual industrial production, greater economic added value and income would be possible by having greater numbers available and employed. A modern economy and high added value is often generated from the service or creative sectors, meaning that if population increase is of mainly low-skilled and low-educated classes, the potential for economic growth, wealth and prosperity as a global aggregate is reduced.

Cultural barriers are another key restraint on economic growth and they are difficult to measure as their impact as a check on growth is neither cause nor effect. Cultural traits in certain urban areas will inhibit economic growth but it will be difficult to say with certainty whether such a trait directly caused such growth to be stifled. Despite this, overcoming cultural barriers may enhance economic growth from a business perspective, especially as wide cultural awareness can impact on any business or organisation wanting to maximise its potential internationally. This economic growth can be realised if staff can deal sensitively and effectively with clients, customers and colleagues from other cultures, particularly if it enables businesses to become more competitive and more profitable. Urban areas with a rich mix of cultural diversity, such as, say, New York City or London, may therefore have a greater potential to trade with other 'non-traditional' cultures and maximise economic growth potential with other culturally different urban markets such as Tokyo or Japan. The example of language is one particular cultural barrier that can demonstrate the point that if there is a strong proportion of multilingual speakers in an urban area, there is a stronger potential to trade and add economic value.

As a further cultural barrier, some cultures may trade in different ways to a western market-based approach that is backed by formal currency arrangements. There may be, instead, a greater reliance on an economic system that operates with personal incentives, not simply in the form of bribes but where economic transactions are enhanced through trust via an emphasis on kinship, favours and friendship. If these particular cultural characteristics are adhered to and accepted, there will be less of a constraint to cross-cultural economic growth.

The fifth key constraint to economic growth is whether a particular space and its inhabitants has access to finance and capital. Financial centres of the globe are almost exclusively located in urban areas, with the top ten financial-centre cities being ranked as New York, London, Tokyo, Hong Kong, Singapore, Shanghai, Paris, Frankfurt, Sydney and Amsterdam (IFCDI, 2011). Some indirect advantages for economic growth in these cities will be enabled due to geographical access and enterprise contacts that work closely with such financial institutions. Global financial capital has a global reach if it can be distributed and generate returns. This means that if particular urban areas can access financial capital, they have the opportunity to invest in the goods and services and hence provide added economic value. Insufficient access to financial capital would therefore inversely restrict economic growth.

The ability to attract global (or domestic) financial capital will be determined by the risk of defaults on debt and its related creditworthiness. This is why debt is seen as a significant constraint on economic growth. For instance, it may be that a nation is spending a large proportion of its economic growth (GDP) on debt repayments and as a result will have a small amount of finance left to invest in projects that could lever further economic growth.

This restraint on growth from financial debt is further stretched and made more complex by the urban areas that will have differing economic strengths and levels of debt. For instance, some cities will have their own local government bank accounts that could be defaulted on or go bankrupt. In some part, local authorities will have guarantees by national central government to ensure outright bankruptcy cannot happen. That is not to say that city administrations have not gone bankrupt in the past. New York City in 1975 was teetering on the edge of bankruptcy and threatening to bring down the entire state. This close call in filing for bankruptcy was due to a declining population since the 1950s due to 'white flight' to the suburbs, a move in industry and commerce to the edges of the city where it was cheaper and easier to operate and the increased social burden from social problems developing, such as high crime rates. From this historical case it can be clearly seen that without control and access to finance, urban areas would have their economic growth stifled. That is unless the expanding and sprawling built environment can financially support any failing pockets or centres of the urban area. Intertwined with this is the ability for the natural resources needed to support such expansion and sprawl being easily obtainable.

The final key restraint on growth involves infrastructure and is slightly less complex than the fluid operations of finance and debt servicing. Infrastructure within an urban area in this sense refers to structures that enable the functioning of society in the built environment. This functioning of society can be enabled through the physical building of basic facilities, services and installations. Examples include transportation and communications systems, water and power lines, and public institutions including schools, post offices and prisons. For urban areas that do not already have such assets, the ability for economic growth will be restrained. For instance, the rapid development of settlements will need infrastructure such as roads, water and power. However, a prosperous and expanding urban area could introduce such infrastructure to loosen such restraints and enable rapid urban economic growth. An example of rapid infrastructure introduction is the Delhi metro system that has provided a highly efficient public transport system. That is compared to the previous heavy reliance on car transport (the highest number of registered cars in any city in the world, at 5.5 million cars) in the city that has caused many million man-hours to be lost each year and the generation of severe traffic pollution.

c. Exploring limits of growth

When considering the internal dynamics of how urban (and national) economies grow, there needs to be some consideration of how far that growth can reach and whether there are actual limits to growth. The debate regarding the limits to growth have been of particular concern in the 1960s and 1970s and have now shifted to arguments, since the 1980s, that centre on sustainable development. That is not to say that a brief consideration of the limits to growth should not be made; especially as, on the one hand, global institutions and national governments still typically argue that only economic growth can provide the resources with which to tackle environmental problems. This is in contrast to many environmentalists' claims that industrial expansion is the root cause of environmental degradation and should therefore be restrained (Cole, 1999).

Several considerations are made here as to whether there are limits to growth. The resources available for extraction from the Earth and its atmosphere limit economic growth. Resources, whether human, natural or capital, may for various reasons provide limitations, although the limit may not be rigid and fixed but more permeable, as human and environmental resources can adapt, reproduce and regenerate. Furthermore, in exploring the limits to growth, it could be argued that some environmental resources can absorb a degree of waste from extraction that is not detrimental to further added economic value.

With regards to the limits to growth due to human resource issues, the current level and growth of population will not only constrain economic growth but also limit growth. A relatively high population in an urban area such as Los Angeles will no doubt put greater strain on the natural environment, with consumption demand for resources such as water, food and energy. This means that the limit to economic growth will be the 'carrying capacity' of an urban area that will provide the minimum standard of living. A minimum standard of living would encompass access and use of the basic goods for human existence such as clean water, food and shelter. Placing a value on a minimum standard of living has been generated in addition to absolute and relative levels of poverty. For the UK, based on views of members of the public, a single person in Britain needs to earn at least £13,900 a year before tax, in 2009, in order to afford a basic but acceptable standard of living (JRF, 2009). This is an example of what economic level, based on individual income, is socially acceptable as a standard of living. But there also needs to be a consideration of the level of economic activity needed to provide the aggregate minimum standard of living for social groups located in space – particularly in urban areas for the purposes of this book; plus an analysis of how much environmental carrying capacity can support this economic activity, especially if there is a limit to what aggregate population level can be supported.

In defining carrying capacity of urban areas, what is meant is the basic infrastructure provisions like water supply and sanitation that help to determine the relative potentials of individual urban centres across regions for sustainable growth (Joardar, 1998). Moreover, consideration of carrying capacity also provides a framework for rational sectoral and spatial allocations of resources for infrastructure development. As these allocations of resources are related to carrying capacity, the urban infrastructure in its basic function of providing water and sanitation, will be the limit to growth at the specific point in time. This 'appropriated carrying capacity', and hence limit, can to be measured, as can other concepts that consider area-based environmentally sustainable concepts. The 'ecological footprint' is one such measure used in common parlance, as well as the per capita 'planetoid', the land available as 'fair earthshare', plus a measured 'sustainability gap' that can demonstrate the changes in consumption patterns that will need to be made to reduce unassimilated, waste seen as an 'ecological deficit' (see Table 5.2). It should be noted, though, that carrying capacity is debatable in terms of being a limit to economic growth if it can be argued that the factors of production are substitutable for one another (Kirchner et al., 1985) and that the carrying capacity is infinitely expandable (Daly, 1986).

Arguing that the level of population as a human resource is a limit to growth is debatable. The strength in the argument that the amount of natural resources places a limit on growth is less contestable. As per the environmental resource economics (ERE) approach to urban and environmental economics, the idea that natural resources are ultimately finite places an absolute limit on economic growth, that is if natural resources are a part of the productive process that generates economic output and cannot be recycled or return to their original form. In the energy sector, the resources used to produce power to light and heat urban areas will be exhaustible if it cannot be replaced or regenerate at a sustainable rate (e.g. oil reserves). Renewable resources, as considered more in ecological economics (EE), begin to add an element that limits, and boundaries for growth are 'fuzzier' if

Table 5.2 Area-based environmentally sustainable concepts

Appropriated carrying capacity	The biophysical resource flows and waste assimilation capacity appropriated per unit of time from global totals by a defined economy or population.
Ecological footprint	The corresponding area of productive land and aquatic ecosystems required to produce the resources used, and to assimilate the waste produced, by a defined population at a specified material standard of living, wherever on Earth that land may be located.
Personal planetoid	The per capita ecological footprint (EFp/N).
Fair earthshare	The amount of ecologically productive land 'available' per capita on Earth, currently about 1.5 hectares (1995). A fair seashare (ecologically productive ocean coastal shelves, upwellings and esturaries divided by total population) is just over 0.5 hectares.
Ecological deficit	The level of resource consumption and waste discharge by a defined economy or population in excess of locally/regionally sustainable natural production and assimilative capacity (also, in spatial terms, the difference between that economy's/ population's ecological footprint and the geographic area it actually occupies).
Sustainability gap	A measure of the decrease in consumption (or the increase in material and economic efficiency) required to eliminate the ecological deficit (can be applied on a regional or global scale).

Source: Adapted from Rees (1996)

resources can regenerate, reproduce or adapt. The use of solar panels to heat and light an urban area would extend the limits for longer-term economic growth, assuming that the cost, energy and resources used to produce the solar panels are economically, socially and environmentally viable. Economics would see a substitution effect taking place from exhaustible to renewable economic output as exhaustible production becomes too costly without incentives from forces such as market investors and government policy.

A particular concept that can consider the more malleable or 'fuzzier' limits to growth is what is known as the 'waste receiving limit to growth'. Here the limits set by both human and natural resource usage may become lengthened or more sustained if the waste generated by such usage can be absorbed. Put simply, if, say, output and waste increases relative to the natural environment's capacity to absorb waste, it will mean that incomes in general can increase up to a point when the effects of environmental degradation start to reduce output and hence income levels. Figure 5.1 demonstrates this trend as a curve that shapes upwards, then flattens out, then takes a downturn, as greater output and waste that needs to be absorbed cannot do so.

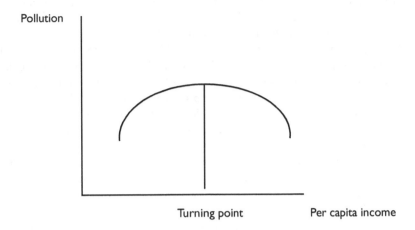

Figure 5.1 An inverted U-shaped relationship between pollution and per capita income
Source: Author

Here, the waste-receiving limit to growth, as measured by income in the example, is at a global level. If such economic growth is to be disaggregated to, say, developed world and developing world, or urban and rural spaces, the limits to growth will start to draw out some complexities and differences. With reference to the developed–developing dichotomy, it is argued that economic development may simply cause a movement of pollution from the developed world to the developing world (Cole, 1999). Complexities will be apparent in urban and rural economic growth differences and how they are played out in terms of environmental pollution limitation. Most prominently, it would be argued that any limits to urban economic growth will be determined by natural resource absorption capacity at a global level as not all growth effects and output will be experienced locally. That said, at a more local level, the growth of a city and its limit from a level of pollution can be made if the pollution type is detrimental to the functioning of a city. Using Mexico City as an example, the very real and visible effects of growth and pollution from industrial production and car usage that is used to access such economic growth is limiting, particularly in terms of the use of the labour factor of production, which will be less efficient and productive if pollution is becoming detrimental to human well-being.

d. Unconstrained growth?

As well as constraints to growth and limits to growth, questions and discussion need to be raised as to whether, inversely, economic growth could potentially break beyond any of the constraints and have drivers that enable 'unconstrained growth'.

One particular driver would be through the developments in technology. If the use of renewable sources of energy and recycling of materials in urban areas were enabled more efficiently through technological developments, it could theoretically provide the potential for unconstrained growth; although counter-arguments will always claim that some resource use will be needed for output. Even with the development of 'zero carbon' homes from technological advance in housing materials, the reality is that energy will be used to produce such materials. The definition of 'zero carbon homes' has yet to be pinned down in the UK (NHBC Foundation, 2009; CLG, 2008a), plus UK government legislation, at 2009, states that the regulatory threshold for zero carbon should be set to cover only those emissions that are within the scope of the building regulations, such as those from heating, ventilation, hot water, fixed lighting and building services – rather than ensuring that builders disclose the lifetime carbon release of a property (Planning Portal, 2011).

Putting aside the effects of CO_2 release, if economic growth is threatened by the exhaustion of resources, the advent of new resource discovery could, theoretically, be one way in which growth could be unconstrained. For example, the discovery of new oil fields or the scientific discovery of new man-made materials, such as insulation material for homes, could extend growth without fear of reaching environmental limitations. The continued promotion and practice of recycling will no doubt generate less waste and provide a more efficient economic system. By keeping more waste within the economic system, it would mean that fewer external costs would be generated, such as the cost of materials only being used once then going to landfill, potentially generating costs of contamination – especially if the land needs to be decontaminated for a different land use that is designated in the future (e.g. residential).

The development of technology has one of the strongest cases in potentially unlocking unconstrained growth. If technology is in advance, the possibility for economic development emerges, and growth can continue unconstrained. If the conditional constraint is environmental resource availability, the advance in less-polluting technologies can shift any perceived boundaries to growth. The advances in renewable energy sources are clearly one step to breaking down any thoughts that exhaustible fossil fuels can halt economic progress. The transfer to renewable forms of energy for consumption in the context of expanding urban areas and urbanisation can be seen as economic growth and wealth continue to increase over the longer term – not forsaking shorter-term global economic growth crashes and uneven patterns of economic development. Substitution to alternative energy production is made as the limited supply (and considerable duty) of oil raises costs to a point where substitute energy products are in demand. The incentive to beat the market by producing cheaper energy is, at present, going to be satisfied by renewable or nuclear sources.

The development of carbon capture technologies may also enable unconstrained economic growth if the threat of CO_2 release as a contributor to climate change is a constraining factor. Carbon capture and storage (CCS) is a mitigation that captures carbon dioxide from fossil-fuel power stations. The CO_2 is transported via pipelines and stored safely offshore in deep underground structures such as depleted oil and gas reservoirs, and deep saline aquifers. This type of development could therefore make mitigation manoeuvres against climate change as more advanced green technologies are enhanced. In practice, in the UK, up to £1bn of capital funding has been made available for the first CCS demonstration project. As such, this is the largest public funding contribution in the world to a single CCS project (DECC, 2011). With regards to resource provision for urban areas, this will make steps in continuing unconstrained growth, assuming that the externality of climate change and global warming adds to the overall economic cost that inhibits growth, particularly if global regulation attempts to halt global warming to less than 2 degrees Celsius above pre-industrial levels by 2050, which will be that much more difficult and costly by up to 70 per cent without CCS (Finkenrath, 2011).

e. Further issues of growth

As mentioned, the direct relationship between economic growth in urban areas and the constraints via the use of environmental resources raise more complex issues that warrant further consideration. The urban issue of unbalanced economic growth in different urban spaces is of particular interest. Some economically booming urban areas may have different degrees of constraints in comparison to other economically declining urban areas. For instance, a city such as Philadelphia, experiencing economic decline in a more developed nation, would not seem to be unconstrained by environmental resource limitations but rather from constraints in economic restructuring, such as increasing human resource capacity by re-skilling a workforce to supply the new service- and creative-sector demands. This would differ to, say, a city such as Lagos in Nigeria, which is rapidly urbanising and experiencing economic growth in a developing nation, which would put more of an emphasis on environmental resource constraints.

Some developing-country urban areas will have economic constraints that are not experienced in more developed urban areas. For instance, unbalanced growth could be generated by the inability of some developing urban areas to have equal access to intellectual property rights such as patents or copyright from ideas and inventions. Resource ownership of this particular property can be held indefinitely over a long period of time and thus maintain wealth by people and organisations in urban locations that are more developed. Further drivers of imbalanced growth for urban areas, even if unconstrained, are that some cities and towns could have access to resources that exclude other urban areas even at national level. The economic growth benefits of cities that are located near an oilfield, for instance, will have economic opportunity at an aggregate scale, although not all residents will benefit individually as an average. For example, in Ankora (Alaska), the city has seen economic growth in the 1970s from oil extraction, with the state's oil revenues paying for almost all state general expenses since 1978, and state spending of those revenues fuelling economic growth, particularly at the peak of production and revenues in the 1980s. The uneven balance of economic growth in cities due to association with certain lucrative commodities can further be demonstrated via this oil case, as, since 1990, in Ankora, economic growth has been slower, because both oil production and state oil revenues have been in decline. As a final point, more complex unbalanced economic growth, even if there were no constraints, is due to the ability of some urban areas to extract wealth and generate environmental waste and be able to export or import some of the constraining factors. For example, a developed nation with urban areas that generate waste will have a greater economic capacity to export waste – in effect dump – to poorer, less-developed nations that are willing to accept a price to import it. Furthermore, some cities will have a better capacity to absorb and recycle resources, such as those urban areas that are sun-rich (e.g. Los Angeles), and will be able to recycle more solar energy as there is more available to process; or those urban areas that are close to water resources and not land-locked (e.g. UK) will have an improved opportunity to recycle their resources for, say, water consumption and energy production.

The issue of imbalances in the unconstrained economic growth of urban areas is further made complex as not all inhabitants of the urban spaces will, proportionately, have economic growth on a per capita basis. The resulting benefits of economic growth in urban areas may go only to a small minority of the population that are part of specific institutions or corporations. For instance, the case mentioned of Lagos in Nigeria may only show a distorted view of per capita wealth, as the GDP contribution of 14 per cent from oil production by multinational companies such as Shell would be held by a few elites who, further still, may not be resident of either the city of Lagos or the country Nigeria.

Political interest may make the idea of unconstrained growth more complex than considering basic limits to growth and sustainable development. For instance, the political capital to gain from appearing 'green' by a political party may cloak the underlying economic and social agendas in order to gain votes and thus power and influence. An example was in the UK, where the current Conservative party, now in Parliament as part of a Liberal Democrat–Conservative coalition, used many references to being 'the greenest party ever'. No doubt this green veneer won support from floating voters, but this approach could still align with its core voters who support an economic policy of regressive tax and conserving wealth among a more affluent class. So political influence and appropriation of power from environmental connections plays a role; and, of course, this is not a limited progression if the ends justify the means. However, if political parties begin to use less socially progressive strategies in the name of environmentalism to gain power, the social outlook will become bleak. For instance, political parties in power may begin to infringe on human rights and civil liberties if the solution to reducing one of the limits to growth is via a 'social cleansing'. Arguments of overpopulation in some urban areas may become a reason to forcefully expatriate residents of a certain social class. For instance, those social groups that are less economically productive (such as the young, elderly and disabled) could be seen (in fascist ideological rhetoric) to be limiting economic growth and consuming scarce resources. This, of course, would be socially unhealthy for the well-being of all society.

Despite the complexities considered in unconstrained growth, a more manageable view can be taken. It is more manageable, for instance, if a more prudent outlook is taken that there are limits to growth, although these boundary limits are malleable and can be changed through the action and interaction of people that manage natural resources. These actions have different timeframes in their effectiveness. The promotion, production and consumption of some goods and services such as energy-efficient technologies are fast responses in potentially extending the limits to growth, as they add to GDP whilst placing less pressure on energy resources. Slower human responses that extend long-term unconstrained economic growth can be explained with regards to changing attitudes to global warming and climate change. A change in consumption patterns by residents in urban areas may take a generation to adopt, but will provide some market signals for a change in more environmentally sustainable or 'green' goods and services.

Summary

1   Economic growth in urban areas will no doubt have consequences in the amount of resources that are being processed. Some of these resources will be extracted from the natural environment and the form of the built environment will also be transformed.

2   A variety of country or regional measures of national income and output to determine whether growth is occurring are used in economics, including gross domestic product (GDP), gross national product (GNP), gross national income (GNI) and net national income (NNI).

3   Negative rates of GDP provide a technical presentation of an economy being in recession. A period where there is neither growth nor a fall in output (or in some cases at a constant low of less than 2 per cent) is where an economy is described as being in stagnation.

4   Output may differ and will have disproportionate consequences for different scales and urban areas, as the fall in economic output will be different in different cities, whilst a national GDP economic growth indicator could be showing a uniform inverse growth rate.

5   The key restraints on growth will have to be explored at differing scales of economic growth (such as the individual to the household to the global) as different restraints will have different magnitudes of control. Key restraints on growth include: natural resources, agriculture, population growth, culture, debt and infrastructure.

6   Limits-to-growth arguments have tended to shift to concerns about sustainability. Some resonating concepts include carrying capacity. Carrying capacity as a limit to economic growth is in debate if the factors of production are substitutable for one another, and that carrying capacity is infinitely expandable.

7   Area-based environmentally sustainable concepts may be more useful contemporary approaches that include carrying capacity, as well as concepts of: an

ecological footprint, personal planetoid, fair earthshare, ecological deficit and sustainability gaps.

8   As per the environmental resource economics (ERE) approach to urban and environmental economics, the idea that natural resources are ultimately finite places an absolute limit on economic growth. Renewable resources, as considered more in ecological economics (EE), begin to add an element that limits, and boundaries for growth are 'fuzzier' if resources can regenerate, reproduce or adapt.

9   In the 'waste-receiving limit to growth', output and waste increases relative to the natural environment's capacity to absorb waste. It will mean that incomes in general can increase up to a point where the effects of environmental degradation start to reduce output and hence income levels.

10 Most prominently, it would be argued that any limits to urban economic growth will be determined by natural resource absorption capacity at a global level as not all growth effects and output will be experienced locally. The limits to growth of a city from pollution will be made if the pollution type is detrimental to the functioning of the city.

11 Drivers that enable 'unconstrained growth' are: (a) through the developments in technology for the use of renewable sources; (b) the advent of new resource discovery; (c) keeping more waste within the economic system; (d) the continued promotion and practice of recycling; (e) the development of carbon-capture technologies.

12 Further issues of growth include unbalanced growth such as: (a) resource ownership imbalances (such as property rights); (b) resource exclusion (e.g. for those without valuable natural resources – land, minerals, water); (c) the ability to export waste; (d) disproportionate wealth and resources per capita (particularly in oil-rich countries); (e) and 'green' politics being used for more social exclusionary ends (such as conservation being confused with conservative, rather than progressive, approaches).

13 Boundary limits to growth are malleable and can be changed through the action and interaction of people who can manage natural resources. These actions have different timeframes in their effectiveness.

Chapter 6

Market forces

Demand and supply

Paris, France

In order to begin to understand the market forces allocating environmental resources in urban spaces, the differentials at both the macro and micro level need to be understood. The theory behind market thinking is also important, as is its use in practice that tends to guide some leading economists and politicians to believe that markets operate best if they are entirely 'free' and efficient. At a more micro-economic level, the focus of goods and services as playing a role in meeting market forces is therefore relevant if elements such as supply and demand are being met in relation to urban areas. The supply and demand of goods and services by urban areas seldom occur entirely free or under direct control. Furthermore, the market contains thousands of producers and consumers, as individuals or organisations, who have different degrees of influence on allocation. It is also worth looking at elasticity of certain goods and services, as this will provide unique insight as to why not all goods and services in urban areas are allocated in a uniform manner.

a. Microeconomics and macroeconomics

Microeconomics and macroeconomics were introduced in Chapter 1, and here this distinction can be explained in relation to markets. To recap, at a smaller level involving individuals and businesses, microeconomics concerns itself more with questions such as the products being produced (e.g. cars), the methods of production (e.g. high-technology production lines), how goods and services are distributed in society (e.g. via the open market) and the efficiency of production and distribution (e.g. is there a high degree of waste output such as CO_2). Macroeconomics is a different sub-branch of economics to microeconomics and asks questions of resource utilisation (e.g. underutilised labour), causes of changes in the purchasing power of money (e.g. inflation) and economic capacity for growth (e.g. natural resource availability).

With regards to markets, microeconomics will aid understanding as to how individuals and businesses interact with the market – particularly in what is supplied and demanded. To use an example of public bus transport as a service in an urban area, microeconomics could help to deconstruct the basic elements of the product, methods of production, distribution and efficiency in, say, the wider market for all transportation 'wants' in a city. Here buses are produced as a good and service in the form of a physical bus that allows residents to move around the city. What buses and how many of them produced will in part be determined by the market, which is determined by suppliers who wish to manufacture and operate the buses, and consumers who will demand a particular bus service at a given price. With regards to the market for buses and the microeconomic methods of production, the manufacture may, for instance, take place in a factory that requires flexible 'made-to-order' methods that ensure that exact numbers required are produced with no surplus stock. With regards to distribution, the market will determine how much customers are willing to pay for the service, possibly given some subsidy from government for those that cannot pay full price (e.g. concessionary rates), as well as depending on the supply cost of producing the required goods for a bus service. Finally, taking a microeconomic analysis of broad market issues involved in an urban bus service, the efficiency of production and distribution may be inefficient to the internal bus business if its seats are not taken up at optimal capacity on every pick-up.

Using the lens of macroeconomics in understanding the role of the aggregate market for urban bus usage, the example can apply aspects of utilisation, purchasing power and capacity. An aggregate market, more theoretically, is one where the price level and real production is used to analyse business cycles, gross production, unemployment, inflation, stabilisation policies and related macroeconomic phenomena. The aggregate market, inspired by the standard market model (as described above in microeconomic analysis of goods and services), but adapted to the macroeconomy, captures the interaction between aggregate demand (all of the buyers) and short- and long-run aggregate supply (all of the sellers). So with regards to utilisation, aggregate market thinking would raise issues as to whether the national aggregate supply of bus services for a nation are being met and bought as an aggregate demand for bus use. Inflation impacts on the aggregate market for bus transport will be those such as analysing if the 'real' price of travel is increasing or decreasing depending on whether the purchasing power of money is rising or falling. For the capacity for growth and the aggregate market for bus travel, macroeconomic analysis would look at whether there is potentially more or less growth in the aggregate market for bus travel in a city than is currently operating.

b. Market forces and economic structures

So markets for goods and services exist in some form and can be used to analyse urban and environmental phenomena via both microeconomics and macroeconomics. To understand how markets operate more deeply, attention to what constitutes the market force should be examined in relation to issues of shared urban spaces and environmental resources. A classical approach to market forces begins with ideas generated by Adam Smith (1776), as introduced in Chapter 1. Here it was argued that the forces in a market are ultimately generated by an 'invisible hand' that operates according to individuals working in their own self-interest, in competition and with supply and demand.

With regards to demand-driven self-interest, key forces are centred on consumer behaviour. Demand-consumption behaviour can arise from changes (or expected changes) in the price and non-price determinants. It is also assumed that consumers are rationally maximising the level of satisfaction (utility) they attain from the consumption of final goods and services. If this individual satisfaction is aggregated to include all consumers, and aggregate demand is high, it will indicate that more goods and services are being demanded, generating a high aggregate level of satisfaction in an economy. This consumer force to satisfy demand is met against another force of supplier self-interest that wishes to maximise its own self-interested satisfaction. Seller self-interest is to earn the highest possible profit from selling goods and services. The seller, or producer, has an objective to make profit by simply ensuring that revenues are greater than costs over a given time period. The profit created over this period is recorded as capital as it can be used to generate further revenue if kept in the business. Note that the level of a producer's profit is indirectly determined via the level of competition in the market.

To apply a microeconomic analysis of a good from the built environment, the example of housing as a consumer good (but noting that housing is also a social and an investment good) demonstrates that the consumer self-interest will be the satisfaction that comes from consuming a house (shelter, cooking, living, sleeping, bathing). From a supplier (or seller) of housing, the satisfaction generated will be the profits made by receiving revenue of a higher price than the purchase (or built) price. In considering competition affecting the level of seller self-interested satisfaction, in a highly competitive market area the opportunity to sell above the average market rate will be lower and less profit will be achieved. The realities of housing supply having unique land-specific location factors make this average market rate difficult to place accurately. This is what is referred to as a good having inelastic supply and will be discussed later and in Chapter 7, when discussing reasons for market failure.

Critique of the metaphor of an invisible hand is apparent, especially in concern over social welfare, as self-interest would have to assume that individuals are altruistically self-interested in providing support for those who cannot support themselves. It is questionable, therefore, whether self-interest can promote welfare in society, if self-interest for more vulnerable members of society is not met by the self-interest of others who wish to see greater opportunity for the well-being of others. The philosophical underpinnings of self-interest are not the focus here, but it is worth noting some critique of the self-interested 'invisible hand' as a force in markets, as the market is seen as a significant force in allocating resources (either environmental or human) to urban areas. One such critique is that the 'invisible hand' is not always there if governments are financing research that the market tends not to put long lead times and finance into. For instance, markets are not good at generating scientific breakthroughs or in providing wider strategic planning, such as in providing physical infrastructure. Furthermore, critique involves issues such as the problem of externalities where a third party has to pay for the market transaction between a buyer and a seller.

Externalities and market failure are covered more in Chapter 7, although the example of environmental pollution is one good example of how market forces are not always working in the self-interest model to maximise social well-being. This also brings into question whether the price, as price is most often attached as a signal in markets, can always be used to measure resource scarcity. What is important for modern economic thinking is the interactive forces operating not in isolation but between each other, especially as the individual forces of the market, the government and the third sector (as defined by community groups and non-governmental organisations) interact to create a position of dynamic equilibrium that is constantly moving and evolving.

The balancing of forces operating in different spheres needs to be unpacked in order to grasp how they interact and are played out in space. The market structure balance in a particular bounded economy at the national administrative area or within an urban area demonstrates some useful context for any further analysis of resource allocation. Different market structures can vary across a spectrum from a completely free market to a completely centrally planned one, with the reality at some point in between. Market economies are characterised by producer decisions such as those made by a business or firm producing goods and services. This is in contrast to centrally planned economies that have most decisions made by a legitimate authority such as a government. At a national administrative scale, the ratio of free-market to centrally planned economies can be quite easily placed. Using examples, nations such as the United States and Japan would be closer to a free-market economic structure, whereas nations such as Cuba, Libya, Myanmar, Belarus, Saudi Arabia and North Korea would be closer to being centrally planned economies.

Mixed economies, such as the United Kingdom – although becoming more free-market as public services are privatised for sale on the market – are the reality for most countries, meaning that for a broad understanding of forces operating, both the free market and centrally planned influence and incentives need to be understood. At this national scale, the broader context of the economic structure will be important in partly affecting how urban areas contained within its borders function.

For urban areas, and especially cities that have a municipal administrative local authority, the role of the market and planning is further made complex. Planning in urban areas will be more 'local' than 'centrally' planned although it does have a non-market role to ensure that its constituents' needs as well as demands (as determined by the market) are met. Within one particular nation, the forces operating within different urban areas will also vary from being more free-market or more locally planned. Some cities will have a strong local-authority power for planning in comparison to others that may have a more *laissez-faire* attitude to promoting business and enterprise. Interestingly, developed world cities that have undergone industrial restructuring since the 1970s now have more changeable attitudes due to the changing economic structures. A city such as Liverpool in the UK, with its loss of population and declining economic growth following the reduction in dockland activity, is a case in point. The political demographic of strong Labour support, as associated with large manual, skilled-labour professions, has meant that support for free-market activity to support regeneration efforts has been met with ideological antagonism but has been enabled through many economic development public–private partnerships (PPPs). Again, economic structures add context at the urban-area scale in order to understand what, how and to whom environmental resources are distributed in urban areas.

c. The market and efficiency

In a market economy, efficiency is the primary criterion to measure institutional performance. As stated above, different economies with different economic structures will provide different conditions in the efficient allocation of resources. If applied to urban areas, there could hypothetically be an understanding of how one urban area operates efficiently in producing fewer waste outputs could be compared to another urban area that is less efficient and produces more waste outputs given a similar amount of resource inputs. More theoretically, economic efficiency is where it is impossible to generate a larger welfare total from the available resources. It is a situation where some people cannot be made better off by reallocating the resources or goods without making others worse off, and thus indicating a balance between benefit and loss. Table 6.1 provides an overview of the five key elements that make goods and services efficient in their allocation. For instance, freedom and choice in the market for, say housing, would make the allocation of housing more efficient in distribution. If there were perfect knowledge of what the housing market was doing in various locations, the allocation of housing would be more efficient in distribution. Furthermore, the increase in economic efficiency would be greater in housing if there was a greater amount of competition between buyers and sellers; a greater mobility in moving house due to technological advances to speed up transactions as well as reduced barriers to selling, such as abolishing HIPS (Home Information Packs); and clearly defined ownership rights of residential property (particularly in the private sector) that enable easy transfer.

Table 6.1 Elements in economic efficiency

Element	Element detail
1. Freedom of choice	Self-interest and rational behaviour.
2. Perfect information	Of market transactions and foresight.
3. Competition	No one buyer or seller can influence trade.
4. Mobility of resources	Changes in consumer preference, income, resource availability and technology are met; no barriers to entry or exit in market.
5. Ownership rights	Clearly defined resource nature and characteristics; exclusive legal ownership rights that are transferable and enforceable.

Source: Author

Part of economic efficiency is the notion of allocative efficiency. Allocative efficiency, also sometimes called social efficiency, means that scarce resources are used in a way that meets the needs of people in a Pareto-optimal way (i.e. at the point of the production possibility boundary – see Chapter 4), and is not to be confused with the concept that resources are used to meet needs as best as possible. To put it another way, allocative efficiency is when resources are allocated in a way that allows the maximum possible net benefit (benefit minus costs) from their use. When an efficient allocation of the resources has been attained, it is impossible to increase the well-being of any one person without harming another person (Pareto-efficient). In applying these elements to the allocative efficiency of natural resources over urban space, a housing theme scenario can be drawn. For housing subsidies, they could be distributed on the supply and/or demand side to aid in providing housing for all citizens. This subsidy will only be allocative-efficient if it meets housing need by providing all with basic shelter. At a point where this provision of housing subsidy is above all housing need, the good of housing begins to lose its allocative efficiency.

d. The demand curve and demand shift

It has been stated that self-interested consumer demand and supply can have a large influence on how resources (natural, human, capital) are distributed in relation to urban spaces. Within the market model, the laws of supply and demand affect the price and quantity of goods and services. In doing so, laws of supply and demand bring together both producers and consumers. In unfettered or perfect markets, these goods and services are produced (the supply) in equal accordance (in equilibrium) to the needs and wants of the consumers (the demand). To interpret market signals perceptibly, for natural resources in the development of urban areas it is necessary to consider the market, or price mechanism, in more detail. In order to explore these markets in more detail, the theory and application of consumer demand is now deconstructed.

The demand for goods and services in the development of urban areas relates to the consumption of those products at a certain price and quantity. As Figure 6.1 shows, the demand curve for most goods and services slopes downward from left to right, as the higher the price, the lower the level of demand (using that economists' term *ceteris paribus* or 'other things being equal'). In the case of demand, product and service examples in the development of urban areas include acquiring land to build on or extract natural resources, recruiting development labour, and using capital goods such as machinery to build physical infrastructure. When economists speak of demand they mean something called 'effective demand'. Effective demand is money-backed desire and is distinct from need; therefore it is demand that can happen because the 'demander' has the resources to pay for it. The determination of demand for goods and services produced by the urban development process is very complicated, partly thanks to the size, cost, longevity and investment nature of the products and partly to the broad range of what constitutes urban development. Discussion of demand in this context is the demand for goods and services in the use and development of urban areas (such as housing, roads, schools, industry, commerce and leisure). These goods and services may not necessarily be sourced from the urban area but will constitute some of the demand for goods and services in urban areas.

Changes in non-price determinants will cause the demand curve to shift to the right or to the left, and demonstrate that more or less is being demanded at each and every price. These changes are often referred to as increases or decreases of demand. It is particularly important to remember the distinction between a movement along, and a shift in, a demand curve. These rules will not only help us to understand the curves, but will also enable us to acknowledge the numerous factors that come into play when interpreting demand in urban and urbanising areas. There are many non-price determinants of demand, such as the cost of financing (interest rates), technological developments, demographic make-up, the season of the year and fashion.

Four major non-price determinants of demand will be briefly focused on here to demonstrate how demand is played out in urban development projects (such as

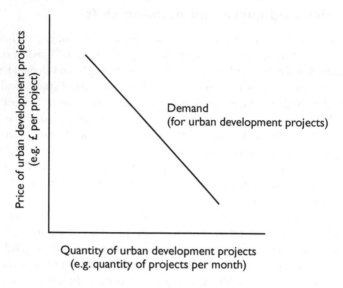

Figure 6.1 The demand for goods and services (as projects) in urban development
Source: Author

the development of residential housing). These determinants are: income, price of other goods, expectations and government policy. First, an increase in income will lead to a rightward shift in the position of the demand curve. Goods where the demand increases when income rises are called 'normal' goods. Most goods are normal in this sense. There is a small number of goods for which demand decreases as incomes increase; these are called inferior goods. For example, the demand for private rented housing tenure may fall if people become able to afford the purchase of owner-occupied tenure homes. Second, the prices of other goods act as a non-price determinant of demand. Demand curves are always plotted on the assumption that the prices of all other commodities are held constant. It may be the case that the other good is a substitute good or a complementary good. In considering these substitutes a change in the price of an interdependent good may affect the demand for a related commodity.

Expectations are a third key non-price determinant of demand. For instance, consumers' views on the future trends of incomes, interest rates and product availability may affect demand. The potential house purchasers who believe that mortgage rates are likely to rise may buy less property at current prices. The demand curve for houses will shift to the left, reflecting the fact that the quantity of properties demanded for purchase at each and every price has reduced as a result of consumer expectations that mortgage rates will rise.

Other non-price determinants of supply include government decisions and policy that can affect the demand for a commodity. For example, reductions in building regulations may increase the demand for housing development, regardless of its present price. The demand curve for housing as part of wider urban development will shift to the right, reflecting the fact that greater quantities of housing units are being demanded at each and every price. If a non-price determinant of demand changes, we can show its effect by moving the entire curve from D to D1 (see Figure 6.2). Therefore, at each and every price, a larger quantity would be demanded than before. For example, at price P the quantity demanded increases from Q to Q1.

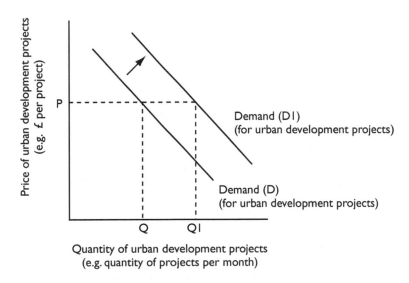

Figure 6.2 A shift in the demand for urban development (e.g. non-price determinants of demand for housing)

Source: Author

e. The supply curve and supply shift

Goods and services for urban areas are made available by producers at a certain price and quantity. Examples of supply for the development of urban areas include land as a site of development, and labour such as tradesmen. Moreover, supply could be the materials required for the actual urban development project, and the capital available could be in the form of machinery and finance. With supply, it is always useful to distinguish between cost and price. Normally, the producer seeks to make a profit, where the cost of the good is less than the selling price.

Economic convention states that the supply curve slopes upwards from left to right, demonstrating that as the price rises the quantity supplied also rises (Figure 6.3). Conversely, as price falls, the quantity supplied falls. As such, the basic law of supply can be stated formally as: the higher the price, the greater the quantity offered for sale; the lower the price, the smaller the quantity offered for sale, all other things being held constant. The law of supply, therefore, tells us that the quantity of a product supplied is positively (directly) related to that product's price, other things being equal. In using an urban development project example of residential house-building, the number of potential contractors interested in bidding to supply a project will increase as the profit margin the client is prepared to accept (and therefore price offered) rises.

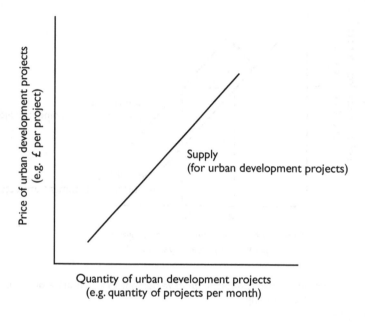

Figure 6.3 The supply of goods and services in urban development
Source: Author

Non-price determinants of supply are important and are demonstrated as a shift in the supply curve. This represents a change in both supply price and supply quantity. Non-price determinants include technology, government policy, supply-chain management and expectations. We shall now broadly consider four of these non-price determinants in turn.

First, if technology improves, the cost of inputs would become cheaper, resulting in a greater quantity of the goods or services to be supplied. For instance, technologically advanced, prefabricated housing production would result in a housing supply that could increase given the same market price, other things being equal. This would be demonstrated diagrammatically as a shift in the supply curve to the right (Figure 6.4). Second, government actions such as taxation and subsidies may create a shift in the supply price and supply quantity. For example, a landfill tax has increased house-building costs and reduced supply at each price. A subsidy would do the opposite by increasing supply at each price, since every producer would be 'paid' a proportion of the cost of each unit by the government. Direct legislation by government can also restrict or release supply in the market and could be in the form of statutory regulations on building, planning or health and safety.

Third, supply-chain management is another non-price determinant of urban development supply. Most urban development activity normally involves integrating and managing many activities to reach the final product. Larger firms and conglomerates can, for instance, subcontract or diversify into other businesses to extend their range of operations. For example, a developer may choose to merge with or take over its material supplier to guarantee that it meets completion targets on time. Such mergers may also reduce supply-chain costs and eliminate many of the associated transaction costs. The final key non-price determinant is expectations, and, more specifically, how expectations about future prices (or prospects for the economy) can also affect a producer's current willingness to supply. For example, developers may withhold from the market part of their recently built stock if they anticipate higher prices in the future. This happened in the 1960s with the Centrepoint building on New Oxford Street in central London and in the 1990s with Exchange Flags in Liverpool. Most other cities can also offer similar examples.

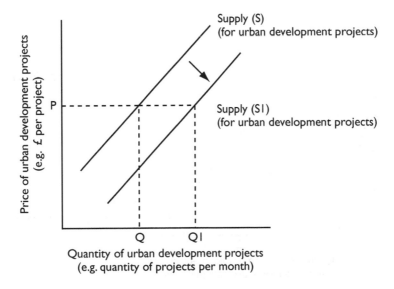

Figure 6.4 A shift in the supply for urban development (e.g. non-price determinants of demand for housing)

Source: Author

More technically, this outward or inward shift in supply can be graphed on a diagram involving the variables of price, quantity and supply. By following along the horizontal axis we can see that this rightward movement represents an increase in the quantity supplied at each and every price. For example, at price P, the quantity supplied increases from Q to Q1. Note that if, on the other hand, the costs of production rise, the quantity supplied would decrease at each and every price and the related supply curve would shift to the left.

f.   Equilibrium: due to deficit and surplus

Now that we have deconstructed some of the laws of demand and supply, it is necessary to explain how both demand and supply interact with each other in a theoretical market place. The classical economic thought underlying the interaction of supply and demand in the market is that they both gravitate towards each other via the force of an 'invisible hand'. The point at which supply intersects with demand is called the equilibrium point. As per Figure 6.5, using the example of a market for urban development projects (such as residential housing developments) it can be seen that the equilibrium point is at the intersection of the range of possible urban development projects demanded and supplied. In reading this market, the market quantity at this equilibrium point is at a quantity of 'Q1' projects (this could be, say, 200 projects started up nationally in one year), and at a price of 'P' unit price (this could be say a price equivalent to a current open-market value of $20 million

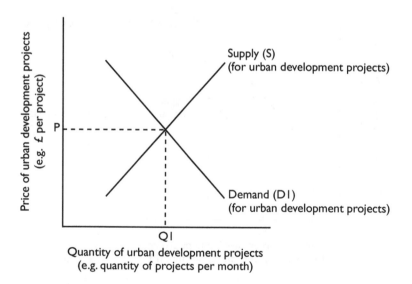

Figure 6.5 Market equilibrium in the supply and demand for urban development projects
Source: Author

for the 200 projects). Classical economic convention argues that the market 'clears' at this equilibrium price at the intersection of demand and supply for all goods and services. This market clearing is not the case for all goods and services, as is explored in Chapter 7 regarding market failure, although the mechanisms that operate to try to make markets gravitate towards equilibrium, if unfettered, are influenced by the occurrence of deficit and surpluses of goods and services. Both deficits and surpluses are now described in order to get a better handle on why the market, in more simplistic economic thinking, tries to gravitate towards equilibrium.

In the case of using a surplus of goods and services in a market, this is a situation where there is an over-supply of a particular good or service if the market is selling at a higher-than-market-equilibrium price (Figure 6.6, Point Qe, Pe). Conceptually, then, it is easiest to think of a surplus in terms of quantity in the first instance, rather than price, and how there is a 'pile' of goods and services stacked up and not being sold at a higher-than-'normal' market price – unsold mountains of corn are a good visual aid. If there is a surplus of quantity, this quantity can be measured diagrammatically as the difference between QS1 and QS2 on Figure 6.6. The higher-than-market price at Ps, rather than Pe, is the price determinant that is related to (but not necessarily the overarching dominant cause of) this excess supply of goods and services.

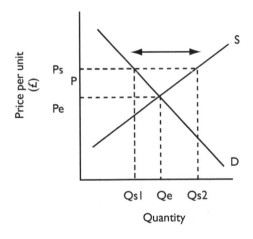

Figure 6.6 A surplus in the market for goods and services
Source: Author

The reaction by suppliers will be to try to reduce the price of the goods and services in order to try to make some revenue rather than letting it perish or rot. Therefore, a revenue and associated profit incentive will force the supplier towards market price. If the supplier does not lower prices, or if there are regulations and tariffs ensuring prices cannot drop below a certain price for a commodity, the surplus quantity could be left to perish or be destroyed. In reducing the price in the market from Ps to Pe, or by reducing the quantity from Qs2 to Qe, the force of movement towards market equilibrium Pe and Qs can be seen. This force of movement to equilibrium can be seen inversely when describing the reaction to a deficit in a market.

A deficit in goods and services is one where there is an under-allocation of quantity at a price that is below the market equilibrium. If deficit is conceptualised, as was considered for surplus, it can be thought of as a lack of physical goods or service or under-supply of quantity; such as a case in which there is not enough corn to feed people in the area that the market can cover and is bounded by. In Figure 6.7, there is a deficit in quantity measured by Qd1 to Qd2, and at a price Pd that is lower than what is the market equilibrium price Pe. In order to react to this deficit in the market, the under-supply of goods and services may be supplied to the market, and as the key determinant is price in this model (non-price deter-minants are not an option here) the result would be to increase the price of the good or service, which will discourage excessive demand for the product and provide greater revenues (price × quantity) and thus greater potential for profit by the supplier. Diagrammatically, Figure 6.7 shows that a supplier will need to increase the price in the market from Pd to Pe or increase the quantity from a deficit Qd2/Qd1 to equilibrium Qe. This could, for instance, be by growing more corn if there is room in this particular market to do so.

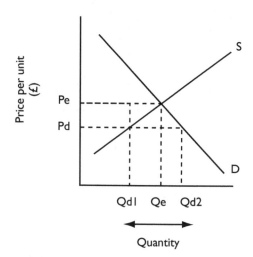

Figure 6.7 A deficit in the market for goods and services
Source: Author

Further environmental resource examples in an urban area when describing deficit and surplus can be applied in addition to the basic food commodity example of corn, although the basic supply-and-demand models for physical built structures, if describing physical urban development projects, begin to become more complex as the goods and services are not as homogenous and transportable as a sack of corn. However, the example of providing more durable environmental resources to urban and urbanising areas can be applied. The environmental resource of iron ore is used for steel production in building materials situations, and can be applied to situations where there is a surplus or deficit. If there was a surplus in steel, due to a slow-down in building property with steel content, the global market price would begin to fall until the price cleared all surplus stocks of steel produced. However, as was witnessed in the 2000s, due to the global property boom, there could be a deficit in steel availability and subsequent increase in its commodity price. This deficit was further geographically skewed in the market towards urban areas that were demanding high volumes of steel for rapidly urbanising areas prior to the property and subsequent financial crash of 2007–08. At a national level, China began to import more steel than it produced, resulting in a trade deficit for China in steel as well as a global deficit in steel availability and high price for the commodity.

g. Elasticity of demand and supply

As well as determinants that shift both price and quantity, economists are often interested in the degree to which supply (or demand) responds to changes in price. The measurement of price responsiveness is termed 'price elasticity'. Price elasticity is defined as a measurement of the degree of responsiveness of demand or supply to a change in price. It is also therefore the ratio by which price changes in relation to changes in quantity, and *vice versa*. In lay terms, elasticity could use an analogy of an elastic band, where a good or service can have greater stretching properties and thus different goods and services can stretch their price to differing degrees depending on how big a quantity of the good or service is increased or decreased.

Demand elasticity

For price elasticity of demand we are therefore observing the magnitude of responsiveness of price demanded in relation to quantity demanded. Furthermore, it can be conceived as the rate of change and deals with the percentage change of quantity demanded in response to a single percentage change point in price demanded. Other variables such as shifts in demand from factors such as changes in income are held constant. In a diagram, the elasticity of demand can be plotted on a graph to demonstrate how elastic the demand is for a particular good or service. As per Figure 6.8, a more inelastic demand (D1) for a particular good or service would be represented by a more vertical line, and thus showing how a small change in quantity demanded will have a price response that is proportionately far greater.

As an example, life-saving medicines such as insulin would be a price-inelastic good due to the quantity demanded having a larger proportional response in price. In essence, people would be willing to pay any price to ensure they received their quantity of life-sustaining medicine. Inelastic demand examples in urban areas include the demand for petrol if there were no available substitutes for car travel (such as a rail service). Here, commuters and in an urban area will be willing to pay a higher-than-proportional price for petrol if the quantity of usage goes up, and, *vice versa*, be willing to maintain consuming the same quantity of petrol given a higher-than-proportional increase in price. In the case of commuting, this is price-inelastic demand for petrol, as consumers will continue to need to travel the same distance to get to their place of employment and earn an income to live – other things remaining equal, such as having an increase in income.

A more price-elastic demand would be a situation where an increase in the quantity demanded for a good or service would result in a less-than-proportional response in the price demanded for the good. As shown in Figure 6.8, diagrammatically this elastic demand would be plotted more horizontally (D2). A demand-elastic example of butter can be used to demonstrate that the price demanded for butter by consumers does not tend to change much as a percentage in proportion to any increase in the quantity of butter consumed. Here the elasticity determinant of substitution takes effect as the consumer can substitute other brands of butter or margarine quite easily.

Where price elasticity of demand for goods and services is more uniform, the demand elasticity is described as being more 'normal' (Do). Any change in the quantity demanded would have an equal and proportionate change in price demanded. The percentage change in fall in price would have an equal negative response in the percentage change in what quantity is consumed. Hence, as per

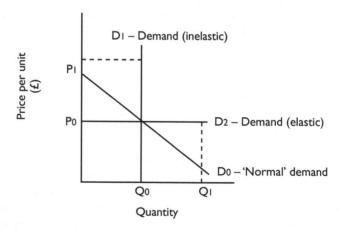

Figure 6.8 Price elasticity of demand
Source: Author

Figure 6.8, the demand curve slopes from top left to bottom right, as there is a negative demand relationship between price demanded and quantity demanded. Examples of 'normal' price-elastic demand would be goods and services that have a price demanded at a negative uniform rate in response to the quantity demanded. For instance, a basic consumer good with 'normal' elasticity of demand would be building materials for property development in urban areas, where it is found that in uniform proportion, the cheaper the price of building materials the greater would be the quantity demanded (other things being equal).

From these examples of differing degrees of elasticity of demand, differing determinants have been present to create the variances in responses between percentage changes in price and percentage changes in quantity. Demand elasticity can be categorised into six determinants, with substitutability being the main over-arching determinant (Table 6.2). The substitutability affecting elasticity is where the price elasticity of demand for goods such as building materials will be more inelastic if there are fewer alternative substitute building materials for development in urban areas. A substitute for steel would have fewer substitutable materials for modern skyscraper construction for instance. With regards to the elasticity determinant of the proportion of income, the demand for a good tends to become more inelastic if the proportion of a good's price is higher in relation to consumer income. The ability to purchase and consume greater quantities of goods with a low proportion of income would mean that the goods are more elastic. The consumption of building materials for residents who wish to make modifications to their property in a more affluent city, containing residents with a greater disposable income, would, for instance, generate a market demand for goods that are more elastic in their characteristics.

Table 6.2 Determinants in the elasticity of demand

Determinant
Substitutability
Proportion of income
Luxuries vs necessities
Dependency
Brand loyalty
Time/duration
Who pays?

Source: Author

Necessity goods are those goods that are essential for daily living and that cannot be cut back on, even when incomes are tight – such as gas, electricity, food and water. This is different to luxury items that are less essential to everyday living. Necessity goods are more price-inelastic in demand, as goods such as food for subsistence, purchased by those residents in an urban area, will be consumed at any price as they are necessary for basic survival. Luxury items consumed in an urban area are less price-responsive to changes in quantity, as goods such as air-conditioning or under-floor heating in a property could not be consumed if incomes are tight.

Dependency and brand loyalty as a determinant of price elasticity of supply occurs because some goods and services will become more inelastic and therefore less responsive to price if a consumer is addicted to an item or loyal to a certain brand. The consumption of substances that have dependency to sustain life (e.g. insulin for diabetics) or loyalty to a brand of retail chains will, equally, make the goods and services more demand-inelastic. The time and duration in which a good or service is held will determine the elasticity, as the longer a person has to think about switching consumption of goods and services based on a price increase, the more likely they are to consume a greater quantity in proportion to price. Using the petrol and urban transport example, if there is a short-term rise in petrol, the decision to use alternative choices in transport such as rail, more fuel-efficient buses, cycling or carpooling would not immediately take place. However, if there was a long period of sustained price rises, the switch to these alternative methods would be more permanent. Petrol in this case would become more demand-elastic over a longer term of sustained high prices.

Short- and long-run supply elasticity

Price elasticity of supply is defined as a measurement of the degree of responsiveness of supply to a change in price. For example, if a 10 per cent increase in price leads to a 1 per cent increase in the quantity supplied, the price elasticity of supply is 0.1. That is a very small response. There are three types of measure that economists use as a reference point to discuss price elasticity:

(a)   Price-inelastic supply occurs when the numerical coefficient of the price elasticity of supply calculation is less than 1. Supply is said to be 'inelastic'. The introductory example in which a 10 per cent increase in price led to a very small response in supply suggests a price-inelastic response: the measured coefficient was 0.1. Normally, price-inelastic goods are those that need to be bought irrespective of the price; an example in urban development would probably be cement. All concrete and mortar needs cement and the majority of buildings use concrete and mortar to some degree.

(b)   Price-elastic supply occurs when the numerical value of the price elasticity of supply calculation is greater than 1. Supply is said to be 'elastic'. For example, a small change in price elicits a large response in supply. This would be an unusual occurrence in the market for property in urban areas, but not impossible.

(c)   Unit-elastic supply is the most hypothetical case, as it describes a situation in which a percentage change in price leads to an identical percentage change in supply. This will always produce a coefficient value of 1. Again, this would be an unusual occurrence in the market for property in urban areas, but not impossible.

In order to improve comprehension of supply elasticity, the use of supply curves can aid by plotting on a graph the supply in relation to measures of price and quantity (Figure 6.9). Price-inelasticity as per point (a) is plotted closer to a vertical straight line (S1), which demonstrates how a change in quantity can generate a greater change in price of the good supplied. Here this vertical line, if perfect, would mean that a change in quantity would have a very large change in the price of the good or service supplied. The example of land as a commodity would be one example attached to having inelastic supply as its fixity to one unique position on the Earth would theoretically mean that the land could be put on the market by a supplier at a far higher price in relation to the quantity increase – this is enabled as there is no substitute land that could be supplied at a lower price.

If elasticity of supply was more elastic, as supply tends to be more in the immediate to short-term, the change in quantity supplied would have a lesser degree of change in the price at which it is sold. Again, this could be due to other substitutes being able to be sold on the market and thus keeping the price change ratio restricted (Figure 6.9, S2). In application, goods with an elastic supply are those that require little capital, no hard-to-find resources and no skilled labour force. It is difficult to find examples of perfectly elastic supply-side goods and services, although one that would aid in differentiating from inelastic supply goods and services is, say, the selling of fruit and vegetables at a market stall. Here, many suppliers can enter the market and there is little opportunity to gain economies of scale. Economies of scale are where suppliers can reduce their per unit costs if the organisation is large enough to gain economic benefits, such as having lower average overheads (heat and light for the factory) per unit as output increases.

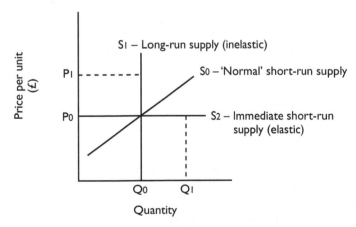

Figure 6.9 Different representations of supply elasticity
Source: Author

Somewhere between supply inelasticity and elasticity is what has been termed 'normal' elasticity of supply. Normal elasticity of supply, often in the short term, is where price changes at an almost uniform rate to the quantity supplied onto the market (S0). An example of normal elastic supplied goods and services are those such as basic consumer goods like fish, which have been supplied from a sustainable farm. Here natural breeding cycles ensure that the supply is not depleted to exhaustion. As a result, a supply of food from this sustainable 'pool' would mean that as a greater quantity is incentivised to be supplied onto the market, the higher sales price and thus profits would be rewarded to the seller.

In the short run, the price movement of physical goods and services (the built environment) in urban areas tends not to affect supply as the increase in supply of any physical urban development good or service takes time. The chief reasons for this are that an urban development project takes time build and urban development projects cannot quickly (if at all) be located to another area of market demand. For physical urban development projects it is common to talk about short-run and long-run supply. The short run is defined as the time period during which full adjustment to price has not yet occurred. The long run is the time period during which firms have been able to adjust fully to the change in price. As an example of residential property in an urban area, in the short run, rental values and house prices are demand-determined because adjustments cannot quickly be made to the supply of property. The markets for residential property development in the short term are therefore price-inelastic in supply. It is this inelastic supply relative to demand that causes property markets to be unstable and characterised by fluctuating prices. In the extreme short run, the supply of buildings or infrastructure is fixed and the supply curve would be a vertical straight line. The example in Figure 6.10 shows how producers supply 2,000 urban development projects no matter what the price. For any percentage change in price, the quantity supplied remains constant (at 2,000 urban development projects) or inelastic. As mentioned, this feature of supply inelasticity is particularly notable within property markets as land is characterised by being perfectly inelastic. This is because land is impossible to move (unless great civil engineering effort is expended, such as developing manmade islands). Existing areas of land can change use and character, but only in the long run as land is developed and property is built.

h. Elasticity examples

Inelastic supply and shift in demand

The interaction of supply and demand curves has been demonstrated to show how in a basic classical economic approach the differing forces interact within the market place. In considering the elasticity of response of price and quantity, a few examples are now given to demonstrate how these variants in elasticity of supply and demand act in the marketplace, particularly when there are changes in the non-price determinants of demand, such as changes in income, and supply, such as advances in technologies for producers. Remember that changes in these

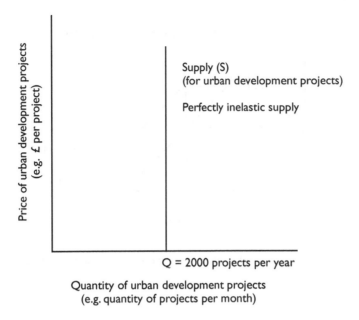

Figure 6.10 Perfectly inelastic supply of urban development projects
Source: Author

non-price determinants will shift the price and quantity, and thus shift the demand and supply curves. First is an analysis of a market where there is an inelastic supply (where a change in quantity supplied onto the market has a greater change in price) and where there is a shift in the demand for the highly inelastic good or service – such as land for building property on.

In Figure 6.11, we can apply the example of the market for food production in urban allotments for consumption by residents in an urban area, holding other things being equal, such as food being allocated to the urban food market from other sources. Here the long-run supply of land (s1) for growing crops in an urban allotment is inelastic as there is only a finite amount of land that is designated in land-use planning for allotments. The initial demand by consumers to grow their own produce in allotments has elasticity considered as 'normal' in that any change in quantity demanded for having an allotment to grow food has a response that is uniform in the price residents are willing to pay for an allotment (D0). Prior to any changes in demand for food grown in allotments the current equilibrium point for producing and growing is at a quantity of Q0 and at a price to rent out or own an allotment space at P0. If there were an increase in the non-price determinants of demand for having an allotment this would shift the demand in both price and quantity from D0 to D1. Shifts in demand for allotment consumption could be an increase in urban population, an increase in income to afford the rent

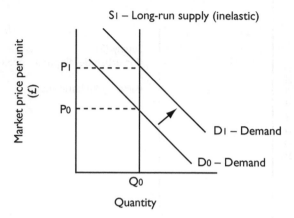

Figure 6.11 The market for urban allotments: inelastic supply and an increase in demand
Source: Author

or ownership of an allotment or a change in preferences to start growing one's own food. The key understanding in this market model is that due to the long-run inelastic supply of land for allotments in urban areas (again, given current land-use planning), any increase in the non-price determinants of demand will have a greater than proportional response in the price of renting an allotment. For urban areas, this may explain why there has been a rise in demand and a lack of supply of allotments.

Inelastic demand and shift in supply

Another illustration is provided to explain how differing elasticity of supply and demand can provide different responses in the changes in price and quantity for a given market. The example provided in Figure 6.12 is with respect to the market for new-build housing. This example considers an urban area where there is an inelastic demand for only new-build housing rather than existing stock. This means that consumers will purchase new-build properties to a large degree at most prices (either very high or very low) as they are within their own personal tastes and preferences at the current time (D0). The supply of housing in this instance at the outset is a 'normal' elastic supply of all housing (S0) that represents the ability of suppliers to uniformly accept price rises at the same rate of change in producing an increase or decrease in the number of homes supplied. Note that the supply of houses in the short term is more inelastic as it would take some months or years for property developers to supply new goods (i.e. new-build houses) onto the market. The initial equilibrium in this housing market for an urban area is at the

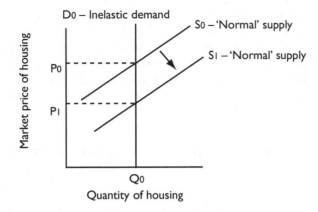

Figure 6.12 Supplying prefabricated homes in urban areas with inelastic demand
Source: Author

price P0 and the quantity Q0. Equating this into real numbers, this could, for instance, be the supply of 500,000 properties at an average price of €150,000.

For the purposes of demonstrating market change with differing elasticity of demand and supply, here an increase in the 'ability' of the market to supply more houses is added to an urban area. Reasons for a shift in the non-price determinants of supply were those such as changes in the price of the factors of production, a greater productivity of the factors of production or an advance in technology for production. In this example, the technological development of housing materials is applied by using prefabricated panels in building homes. With regards to the market, the increase in technological production methods 'enables' (not necessarily actually supplies) a greater quantity of homes to be built at all price points compared to the older technological methods. This non-price determinant increase in supply shifts the curve to the right to generate a new equilibrium point (P1 and Q0) from the original point in time (P0 and Q0) prior to the technological advances.

The key point here in the case of inelastic demand for housing (where in this instance home purchasers only want new-build property types) and shift in supply of housing due to technological advances is that the quantity of housing (Q0) being built stays almost the same whilst there is a greater than proportional fall in price (P0 to P1). This large percentage change fall in market price for housing due to technological advances would result in cheaper building costs and a resulting reduction in revenues and profits for those selling the properties in his market (revenue change from P0 × Q0 to P1 × QO). In short, quantity of demand for new-build has stayed the same, as people only want to purchase new-build rather than substitute for older stock. The potential to supply has increased by a non-price determined factor of technological advances. This has played out in the market-place as a fall in overall market value, as it is cheaper to put technologically advanced housing products onto the market. This example is obviously simple at the outset but demonstrates some of the market forces at work for a consumption good. Further analysis of the housing market will realise that the 'good' of housing is made complex as it is a consumption good (bought and sold), a social good (supported by the government as a basic human need) and an investment good (influenced by changes in financial circumstances).

Petrol and tax allocation (demand inelasticity matters)

The operation of elasticity of demand and supply in markets can further be applied to examples in urban and environmental economics. Within Figure 6.13, the example used is of the taxation on petrol, which could, arguably, be an action taken by an authority to deal with the costs of pollution in oil extraction, distribution and consumption that will increase the release of CO_2. Further reasons for taxation by government will be to use the revenues from natural resource depletion and real-locate funds towards other public goods. Public goods are those goods that have no rivalry and are non-exclusionary in society. The free market cannot alone provide public goods, such as defence (military) and many environmental goods such as clean air.

The market characteristics of petrol need to be understood in order to see how the producer and consumer would meet the incidence of a tax. Petrol is a substance that has an inelastic demand in that if the quantity of petrol demanded increases there will be a greater than proportionate change in price willing to be paid in the short term. People will still consume petrol at higher prices as there are not many substitutes for petrol in fuelling cars at present; although over the longer term this may become more elastic as people have the option to switch to hybrid biofuel or electric cars. Petrol supply in the model shown in Figure 6.13 is seen to have a 'normal' elasticity (S0 and S1), in that suppliers of petrol will experience a uniform rate of price rise if the quantity of output is increased. If 100,000 barrels are supplied for a price of $1 million, an increase in supply to 1,000,000 barrels would result in a uniform price rise with sellers willing to charge at $10 million. The effect of a tax would be to constrict supply as it is a non-price determinant that can be attached directly to the supply of petrol.

The inelastic demand for petrol (no immediate substitutes for petrol in cars) means that the constriction in supply by producers from a tax on the good will result in different proportions of tax incidence paid by the consumer and producer. The consumer will pay the largest proportion in tax, as the new market equilibrium-clearing price moves from E0 to E1. As the new equilibrium creates a price rise (P0 to P1) to recoup the quantity lost (Q0 to Q1) in movement *along* the demand curve (E0 to E1), this loss in revenue will be what is paid for in tax should the quantity supplied remain the same as originally supplied onto the market. The incidence paid by the producer is due to the new equilibrium (E0 to

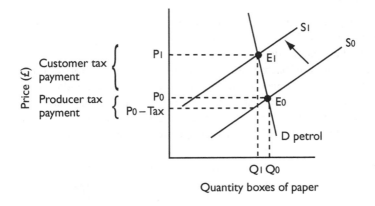

Figure 6.13 Petrol taxation example: Who pays pollution tax?
Source: Author

E1) generated through the supply curve *shift*, and means that the producer has to pay tax on each unit sold ('P0' to 'P0 − tax') if they wish to maintain supplying at the original market quantity with a tax imposed.

The implications from this example are that larger proportions of the tax burden are incurred by the consumers of petrol within and travelling to urban areas, and thus this imbalance of burden can be held as a significant factor when managing environmental resources through economic means. For instance, this model demonstrates that a simple increase in tax on petrol will not necessarily change consumer behaviour as they will continue to purchase the good, and pay a large proportion in tax. The lower proportion of the tax burden incurred by the suppliers of oil means that continued high profits can be made without much contribution to tax, and therefore they may continue to supply petrol with not much price disincentive to invest in selling alternative energy goods. It is noted also that this model centres on only a handful of variables that change − price, quantity, supply, demand, and the imposition of tax on the good being supplied. Other wider factors will have to be considered such as setting each market for petrol in context as different geographies will have different markets; the local market for petrol in London will be different to the local market in Abu Dhabi, as it will also differ from how the global market for petrol operates. The temporal context is also important as short-term and long-term market trends will show different patterns, and the market will respond differently whether it is at a relative peak or a relative trough. Also, the operations of the market are not just economic rational price–quantity demand–supply decisions. The impact of political lobbying and decisions will affect the market, as will educational influence on increasing awareness of urban and environmental issues that, in turn, can affect market behaviour.

Summary

1   With regards to markets, microeconomics will aid understanding as to how individuals and businesses interact with the market, particularly in what is supplied and demanded.

2   Using the lens of macroeconomics in understanding the role of the aggregate market can apply aspects of utilisation, purchasing power and capacity.

3   A classical approach to market forces begins with ideas generated from Adam Smith (1776). Here it was argued that the forces in a market are ultimately generated by an 'invisible hand' that operates according to individuals working in their own self-interest.

4   The consumer force to satisfy demand is met against another force of supplier self-interest that also wishes to maximise its own self-interested satisfaction. Seller self-interest is to earn the highest possible profit from selling goods and services. The profit created over this time period is recorded as capital as it can be used to generate further revenue if kept in the entity.

5   Critique of the metaphor of an invisible hand is apparent, especially in concern over social welfare, as self-interest would have to assume that individuals are altruistically self-interested in providing support for those who cannot support themselves.

6   Different market structures can vary across a spectrum from completely free market to completely centrally planned, with the reality at some point in between. At this national scale, the broader context of the economic structure will be important in partly affecting how urban areas contained within its borders function.

7   For urban areas and especially cities that have a municipal administrative local authority, the role of the market and planning is further made complex. Planning in urban areas will be more 'local' than 'centrally' planned, although it does have a non-market role to ensure that its constituents' needs as well as demands (as determined by the market) are met.

8   Some urban areas will have a strong local authority power for planning in comparison to others that may have a more *laissez faire* attitude to promoting business and enterprise.

9   In a market economy efficiency, is the primary criterion to measure institutional performance. Allocative efficiency is where it is impossible to generate a larger welfare total from the available resources. It is a situation where some people cannot be made better off by reallocating the resources or goods, without making others worse off and thus indicating a balance between benefit and loss. Five key elements that make goods and services efficient in their allocation are: freedom of choice; perfect information; competition; mobility of resources; and ownership rights.

10   In unfettered or perfect markets, goods and services are produced (the supply) in equal accordance (in equilibrium) to the needs and wants of the consumers (the demand).

11   The demand for goods and services in the development of urban areas relates to the consumption of those products at a certain price and quantity. The demand curve for most goods and services slopes downward from left to right, as the higher the price the lower the level of demand.

12   Demand as 'effective demand' is money-backed desire and is distinct from need; therefore it is a demand that can happen because the 'demander' has the resources to pay for it.

13   There are many non-price determinants of demand (a shift in the demand curve) such as the cost of financing (interest rates), technological developments, population and demographic make-up, the season of the year and fashion; income; price of other goods; expectations; and government policy.

14   The supply of goods and services for urban areas is made available by producers at a certain price and quantity. Examples of supply for the development of urban areas include land as a site for development, and labour, such as tradesmen. Supply could be the materials required for the actual urban development project, and the capital available could be in the form of machinery and finance.

15   The law of supply tells us that the quantity of a product supplied is positively (directly) related to that product's price, other things being equal. Non-price determinants (a shift in the supply curve) include technology, government policy, supply-chain management and expectations.

16   The classical economic thought underlying the interaction of supply and demand in the market is that they both gravitate towards each other. The point at which supply intersects with demand is called the equilibrium point.

17   A surplus of goods and services in a market is a situation where there is an over-supply in quantity of a particular good or service, if the market is selling (rather than buying) at a higher-than-market equilibrium price. A deficit in goods and services is one where there is an under-supply (rather than demand) of quantity at a price that is below the market equilibrium.

18   As well as determinants that shift both price and quantity, economists are often interested in the degree to which supply (or demand) responds to changes in price. The measurement of price responsiveness is termed 'price elasticity'.

19   A more price-elastic demand would be a situation where an increase in the quantity demanded of a good or service would result in a less-than-proportional response in the price demanded for the good. Demand elasticity can be categorised into six main determinants. These are: (1) substitutability; (2) the proportion of income; (3) luxuries vs necessities; (4) dependency and brand loyalty; (5) duration; and (6) who pays.

20   If elasticity of supply was more elastic, as supply tends to be more in the immediate to short term, the change in quantity supplied would have a lesser degree of change in the price at which sold. Again, this could be due to other substitutes being able to be sold on the market, thus keeping the price-change ratio restricted.

21   In the short run, the price movement of physical goods and services (the built

environment) in urban areas tends not to affect supply as the increase in supply of any physical urban development good or service takes time. The main reasons for this are that an urban development project takes time to build and that urban development projects cannot quickly (if at all) be located to another area of market need. For physical urban development projects it is common to talk about short-run and long-run supply.

22 Elasticity examples can demonstrate how markets interact and operate with (a) a shift in demand for allotment space in urban areas that have short run inelastic supply; (b) the case of inelastic demand for housing (where in this instance home purchasers only want new-build property types) and shift in supply of housing due to technological advances; and (c) the larger incidence of consumer tax on petrol due to a more inelastic demand for the good.

Chapter 7

Failure of the market and externalities

Boston, USA

This chapter will begin to move beyond basic neoclassical market attempts at understanding how goods and services (particularly environmental products) in relation to urban areas are consumed, produced and allocated. Examples of regulating for public goods such as clean air and a more sustainable climate can be dealt with, economically, by internalising the problem; such as in the case of pollution by firms, by ensuring either party as consumer or producer in the market transaction pay for the cost generated. It is considered here that markets may inherently have what is termed 'market failure' in the way in which some markets can fail in distributing resources efficiently for all of society.

Furthermore, it is demonstrated here that all social and economic costs cannot necessarily be internalised and met by the two agents, producer and consumer, in the market; there is a third external party that will incur a cost that will not always be met by the consumer or producer. This is where the concept of externalities is introduced. With regards to the subject of this textbook, applications of market failure and externalities are made to urban and environmental phenomena. The long-term over-supply as a result of low demand for housing in many cities can be seen as an inefficient allocation of resources for housing and therefore a sign of market failure in the housing market. Illustrations of externalities are those such as pollution and CO_2 release that are symptomatic of the market failing to internalise payments for all costs in manufacture and service provision of environmental products.

a. What is a market? revisited

In order to have a clearer account of market failure as well as the economic phenomena external to the market, an understanding of what a market is in the first instance should be reaffirmed. A market is defined here as the place in which a free transaction of two parties takes place – nominally a buyer and a seller – who will agree to exchange a quantity of goods and services at an agreed price. This agreed price in the marketplace is what is termed the 'market clearing price' and will be determined by forces that operate in the self-interest of both the buyer and the seller. Markets, in reality, are not entirely 'free' in that the agreed price and quantity exchanged at, will be influenced in some way by market imperfections and guidance from authorities that can guide and control the force of an 'invisible hand'.

At a broader macroeconomic level, countries, and indeed urban areas, will have differing balances in the forces that shape free-market activity. The economic context in which resources are allocated will depend on the positioning of an area in a spectrum that will range from a completely free market economy to one that is centrally planned. Those areas that are set within a freer market context will have decisions in allocation shaped by self-interested consumers and sellers within a more private transaction, whereas urban areas set in a more centrally planned economy have goods allocated more on the grounds of government rules and regulations. A primary objective for producers, for instance, is to sell goods as profit maximisers, where the revenue income is higher than the costs paid out over a designated timeframe. Government interest in the market may depend on the party in power at the time, in whether they are in favour of supporting a freer market approach to the economy – and thus influencing the urban areas contained within its national boundaries. More broadly, from a purely political outlook, the government is in power to serve the interests of the public, which is the interest of every member in a given society, no matter what ability to pay each member of society has. The reality is far from this simple duality of market contexts, although it enables a good starting point to realise differing forces that influence market activity to either encourage or rectify market failure and externalities. For example, a regeneration and renewal project using both private and public funds, in a PPP (public–private partnership) is difficult to clearly segment into ownership as the nature of the contract may shift ownership as the project develops.

b. Revenue and cost analysis

So far we have introduced the most basic of economic tools, the analysis of markets using graphs to plot and help to explain how demand and supply interact with each other. As seen previously, the self-interested forces of demand and supply meet at an equilibrium point at a market clearing price of P1 and a market clearing quantity of Q1 (Figure 6.5). As discussed in Chapter 6, the upward slope of the supply curve represents how a supplier will be incentivised to sell more goods and services if the price per unit sold in a market increases. Underlying this principle of selling more as price per unit increases is that of the profit maximisation motive of the seller. Profits are determined by subtracting the costs of producing a good or service from the revenues gained when selling to a buyer. The revenue will be the multiplication of the price sold by the quantity of units sold.

With regards to profit maximisation as an underlying principle, it is defined as a process that organisations undertake to determine the best output and price levels in order to maximise return. To reach a maximised level of profit, an organisation will usually adjust influential factors such as production costs, sale prices and output. There are two main profit-maximisation methods used, and they are the marginal cost–marginal revenue method, and the total cost–total revenue method. Marginal cost is the cost of producing one more unit of a good or service, whereas marginal revenue is the income generated by selling one more unit of a good. Therefore, the marginal cost–marginal revenue method of profit maximisation is the point at which the organisation will make zero profit if any more additional units of goods or services are produced. Under the total cost–total revenue approach, the organisations profit is maximised at the output level when the total aggregated costs equal the total aggregated revenue obtained from sales, and hence no more profit can be obtained from producing more output of goods or services.

c. Cost of 'free' environmental resources

Costs are therefore intrinsic to price and quantity in the supply of goods and services onto the market. If it is assumed that firms operate rationally as profit maximising, the incentive to conserve costs is strong and a great deal of attention will be placed on reducing the cost base or, more colloquially, the 'bottom line'. In economics, the costs internal to the firm and the costs external to wider society and the environment are of importance, particularly as transactions will have a cost imposed on a third party not involved in the internal market transaction. Both the internal cost and external cost are therefore referred to as the full cost in producing a good or service. As an equation, if the full cost can be quantified, it is expressed quite simply as:

- Full cost = *internal* cost (firm) + *external cost* (social and environmental)

Some examples can be applied of environmental resources that are used in production within urban areas and to be consumed by urban dwellers. Firstly, the prominent issue of energy needs by urban populations means that energy firms are significant contributors to both internal and external costs. For a nuclear power plant, for instance, the internal costs will be both direct and indirect. Direct costs internal to the firm will be those attributable to the production of energy such as the uranium used in the nuclear process to produce energy. Indirect internal costs for the nuclear energy firm are those not directly attributable to the production of energy, such as heat and light overheads for the operations of the plant. Some of the direct costs will be external costs and these are costs that are incurred by a third party, such as those communities incurring a cost even though they have not benefitted from the production or consumption of the energy good.

An external cost could be the environmental cost paid by society at large through ill health of those people exposed to radiation and the subsequent knock-on costs of a degraded environment (e.g. radiation in the food chain). The cost of storage of nuclear waste will be a direct external cost to a third party and therefore needs to be paid for and internalised by the firm, otherwise the social and environmental cost of unsecured radioactive waste disposal will have been avoided by the producer or consumer. To place a cost and internalise the external cost of safe disposal, transportation and storage, in the United States a surcharge of a tenth of a cent per kilowatt-hour is added to electricity bills. Hence the consumer in the market place meets the external cost in the market place for nuclear energy. As a formula, the internal and external costs of energy production as a full cost are:

Full energy cost = *internal* electricity processing cost + *external* environmental cost

It is important to note that within the economics of resources that have differing degrees of environmental consequences in extraction, it is more difficult to attribute a full price because the full cost is not always internalised, particularly, as the selling price is often generated by adding a mark-up (the alternative could be a margin) to the costs of production. The result of not internalising an environmental resource cost that is difficult to price is a situation of market failure. As previously discussed, a situation of market failure is where the allocation of resources is not efficiently provided within a 'money-backed' marketplace, such as the private market for energy supplying the energy demands of an urban population at an inefficient price and quantity. If the external pollution costs of radiation are not within the market price, some other party in society will be inefficiently paying for costs that they have not derived any utility or satisfaction from, Turner *et al.* (1994) describe this market failure due to environmental resources being 'unpriced' and therefore incapable of being internalised into the marketplace. It is stated that:

> while the market system appears to be highly efficient at using priced resources, it fails to correctly guide firms towards the efficient use of unpriced environmental resources. This Market Failure arises because firms only take account of the market price of a resource when deciding how much of that resource to use.

d. Externalities

What we are discussing in relation to surplus external costs is, in essence, what economists refer to as 'externalities'. It is worth here explaining in more depth what externalities are and how they feature within urban and environmental considerations. Externalities can be more broadly defined as the spill-over or third-party effects arising from the production and/or consumption of goods and services in the marketplace. Transactions within the market will generate costs and benefits that are external to those producing and consuming goods and services. For instance, a supplier (e.g. a construction company) provides for purchase on the market a number of buy-to-let residential properties within a city neighbourhood. In this transaction a building supplier will sell these units at the maximum possible price assuming the seller is rational and wealth-maximising. Similarly, the buyers (the new landlords) in the transaction will pay for the units at the lowest possible price negotiated, again this is assuming the buyers are wealth-maximising and rational. Under normal circumstances there will be benefits gained by both parties, although a transaction may be at a loss under certain circumstances, such as if there needs to be a quick sale or if the market price for residential property is falling.

What is not factored in within this market transaction is what the external costs and benefits will be for those other households in the neighbourhood that will interact with and gain satisfaction from proximity to the new dwellings. As examples in this case of negative externalities, there could be the problems (e.g. noise, fly-tipping, poor maintenance) of transient families in buy-to-let properties that have no long-term stake in the neighbourhood. Poor upkeep of homes by residents with no incentive to 'keep up appearances' may, in turn, devalue the neighbourhood. It is this fall in value that is not internalised in the original market value and hence is considered as having a measureable external cost, referred to in short as an externality.

Externalities can be positive and/or negative with a third-party spill-over effect generating benefits as well as costs. In using the housing market example, a positive externality could be if the buy-to-let properties were providing a new higher-income demographic of resident to a neighbourhood that was depopulating and of low value. The introduction of a new incoming group of residents to a neighbourhood that was depopulating or was suffering problems of large-scale concentrated mono-tenure (e.g. social housing) may provide some external benefits. For instance, the spill-over effects of increased population and increased disposable income in a struggling neighbourhood will enable local services to continue to have a demand. Furthermore, the introduction of a tenure mix will encourage the development of a mixed community that will provide opportunities for greater social cohesion and reduce social polarisation of communities in neighbourhoods at the city scale.

To use a more simple physical, rather than social, example of positive externalities in a neighbourhood, the painting of the exterior of houses will have generated costs and benefits that are both internal and external. For the seller, the internal costs would be the cost of the resources used to make and sell the paint, plus the cost of the paint. For the buyer it will be the final retail cost of the paint and the labour cost in painting the houses. The internal benefits to the supplier of paint will be those such as rewards in profit, and the buyer will internally benefit by improving the aesthetic of the house that can increase value. With regards to externalities, the external costs 'spilled over' to a third party is if the painting is of a poor quality or the wrong colour, and is not in keeping with its environment, having a detrimental effect on the value of a neighbourhood. If the painting of properties improved the neighbourhood environment, this would generate a positive externality. The use of cost-benefit analysis is one technique explained in Chapter 8 that enables analysis of all costs and benefits in a project (e.g. the painting of a house, or the building of a bypass) to be identified, valued and subtracted from each other to produce a final net benefit or cost.

e. Failure to achieve efficiency and other social goals

It has so far been considered that markets can be conceptualised, although the workings of the market are not as clear and simple as classical economic market theory would suggest. As described above, there are limitations of not being able to internalise all costs and benefits, and third-party external costs and benefits need to be recognised outside of the market transaction. Further market difficulties are their failure to efficiently allocate resources, whereas conventional free-market theory would suggest that equilibrium of supply and demand is always achieved. 'Market failure' is the key concept for discussion here, and is where the market fails to achieve an efficient allocation of resources. This inefficient allocation of resources, certainly applies to environmental resources in their use, production and consumption within urban areas. Illustrations of these inefficiencies will now be made with reference to four significant reasons for market failure.

Table 7.1 demonstrates the five main reasons why markets fail, and this is recognising that markets exist as a means of resource allocation. The occurrence of monopolies are one way in which the efficiency of markets is restricted, as one supplier in the market would mean that the supplier would be able to fix prices at a higher rate due to lack of competition. Lack of competition may be because there

are barriers to entry into the market; for instance, the telecommunications infrastructure in urban areas (and beyond to national boundaries) has often been
monopolised by companies such as AT&T in the US and British Telecom in the
UK. This monopoly power and domination in the telecommunications market has
meant that prices are held artificially high without the ability of competitors to
enter the market. This has prompted governments to step in and ensure that
companies act in the interest of consumers by breaking into smaller de-merged
entities with a lower percentage of market share. Monopolies also occur for very
practical reasons such as the strategic benefits of being a natural monopoly. A natural monopoly is where it makes sense to have only one producer in a certain
market, especially if the addition of one extra unit means that the unit cost is lower.
An example of a natural monopoly within an urban area is the utilities that service properties such as electricity, gas, water and sewage. It would not make
economic sense in terms of efficiency if several companies built these service infra-
structures, as this would mean that there would be four or five cables and pipes
servicing properties in a city when only one would suffice. The practical economic
solution would be to have a well-regulated infrastructure that allows service
providers to lease and compete for the single physical infrastructure. Transport links
between urban spaces could also be regarded as forming a natural monopoly and
hence demonstrate that a complete free-market approach would be inefficient. The
rail network that connects cities within national boundaries is often owned by a
regulated co-ordinating body who lease out parts of the track to competitive
tender. For the UK, the selling off of state-owned British Rail generated a situation where passenger and freight train operators can lease or hire stock.

Secondly, the occurrences of external costs in addition to costs internal to the
market are a reason why markets are deemed to be inefficient and therefore fail.
Externalities have been discussed in the previous section and refer to the third-
party spill-over effects that will either cost or benefit those that are not directly
engaged in the market transaction between seller and buyer. The urban and environmental example used was the effect on the community from the buying and
selling of property in a neighbourhood. For instance, the building development and
sale of properties designed mainly for rent may encourage a more transient resident
base and potentially affect the characteristics of a community, or potentially devalue
a neighbourhood stock of property. What externalities can demonstrate is that
external costs, such as poor maintenance of rental properties, are not considered
within the private market sale, and the spill-over costs are not integrated into the
full economic cost.

The presence of public goods, as well as private goods, is another reason why
markets are deemed to be inefficient and fail. A public good exists when a person
cannot be excluded from its provision and when one person's consumption of a
good does not reduce its availability to anyone else. Public goods therefore have
two conditions of non-excludability and non-rival consumption, whereas private
goods have two conditions that are excludable and have rivals. As an example, services that try to protect people from climate change or protect biodiversity will be

classified as a public good. In these examples, they are non-exclusive as no being can be excluded from the benefits of CO_2 reduction and climate change, and they are non-rival as the marginal costs incurred from promoting a biodiverse ecosystem will not increase because no rivals can benefit (economically) from entering the market.

The potential problem with using the market to provide this public good voluntarily is free-riding, since he or she cannot be excluded from the same amount of the good. The presence of free-riding means that the market cannot efficiently incentivise the quantity of consumption of a good or service by changing price – as some consumers will always be able to consume at zero price. For instance, parks in urban areas (or more technically, urban green space) are a sought-after public good and are an aesthetic available to all people in the same amount. They are non-exclusive and non-rivalry as one person's visual consumption of open space does not reduce another person's access nor reduce the amount of open space. The payment to local and central government in the form of taxes (or as public assets accumulated over many years in taxes and other revenue income) to maintain and keep such parks as public goods will mean that those not contributing to the tax base can 'free-ride' the cost but gain in its benefits.

The market fails in the case of public good provision because people will undersupply the socially optimal level of a public good. This is because in the private market, people can contribute an amount less then their true benefits for the good. To use the urban green space example, the provision of a park that is available to all may not be provided efficiently by the market, because if the market supplied a park, different contributions would be volunteered by consumers as the benefits from the park would be gained even if a zero amount was offered to the market for a park.

Some resources that have ownership attached to them may be common in that they are owned by no-one and used by anyone. Someone owned parks in the previous example used, in that they were owned by the public sector. A common ownership of property resources, and those owned by no-one and used by everyone will mean that the market cannot efficiently provide such resources. This is because common ownership means that it cannot easily be turned into a commodity and sold in quantifiable units. An obvious example is the clean air that is desired to be breathed by all people. The market cannot assure that all people can have access to clean air, whilst it simultaneously can assure clean air for those that do not pay for their environmental location. This means that the market fails in efficient provision of a common property resource such as air (the same might be said of common forests, or of the high seas for fishing), particularly as less-mobile people cannot easily access this good economically.

In a similar vein to common property resources, if there are weak property rights for some goods and services, it can mean that they are not efficiently allocated in the market system. Most of the market failures with environmental assets can be linked to incomplete markets. Markets are incomplete because of the failure or inability of institutions to establish well-defined property rights. For example, many people own land and are able to take action when damage is done to it, but they do not generally own the rivers or the air, through which a significant amount of pollution travels. The lack of clear and well-defined property rights for clean air thus makes it difficult for a market to exist such that people who live downwind from a coal-fired power plant can halt the harm that the plant does to them or successfully demand a fee for the costs that they incur.

The fifth factor that explains failure in the market is that of asymmetric information. Asymmetric information is a situation where all the consumers and producers do not have the same information on which to base their decisions. The implication of asymmetric information in the allocation of environmental resources to urban areas is that a regulator of pollution levels by industry in proximity to urban areas will not have a perfect reading of how much pollution is being emitted. This gap in actual and recorded value therefore provides some element of a moral hazard in which regulators can over- or under-identify the 'official' level of pollution. This gap will guide the market in terms of directing pollution costs to be met and thus provide a market allocation that is not efficient if the information is inaccurate or different to what those involved in the market were expecting. As well as the moral hazard, there may also be an adverse selection if there is asymmetric information. An adverse selection could be that the polluters in a market influence regulators to record lower levels of the unknown pollution reading, as this will reduce market costs. These lower market costs by firms will therefore be different to what costs should have been paid to account for pollution control – and therefore not at the more efficient full market cost.

Table 7.1 Types of market failure and descriptors

Type of failure	Descriptors
Monopolies (including natural monopoly)	• Few producers relative to the size of the market
Externalities	• External social and environmental costs, in addition to internal market costs
Public goods (and common property resources)	• No exclusive and non-rival • Owned by no-one, used by anyone
Weak property rights	• Inability of institutions to establish well-defined property rights – market doesn't function in a complete rational and bounded process
Asymmetric information	• Producers do not have the same information • Generates moral hazards and adverse selection

Source: Author

As well as the five basic features of market failure, the market also fails to achieve other socioeconomic goals. These socioeconomic goals are particularly pronounced in urban areas where a more-dense population interact both socially and as isolated individuals. Three key goals are stressed and these involve distribution, protection and obligations. Failure to achieve distribution is how the market can fail to achieve an equitable income distribution, especially for those social groups such as the disabled and elderly that do not have the labour capacity to generate wealth for themselves without assistance. Without social capital derived from family or other community networks the redistribution of income or support from public funds would disadvantage such groups. It is these redistribution mechanisms from family, community and public sources that the market will not necessarily allocate.

Failure of the market to achieve the social goal of protecting individuals is also apparent. A civil society has developed to strive for individuals to be protected from harm by others or themselves. The price mechanism will not necessarily protect as a form of paternalism from harm and external forces will need to intervene if this socially optimum state of well-being is preferred. The compulsory use of seatbelts is often cited to demonstrate how regulation can ensure members of society are protected from harm to themselves and others when the market would fail to do so. Remember that failure in economics is the market not efficiently allocating resources, and in the case of seatbelts, unnecessary death is not economically efficient for a harmoniously functioning society. A more direct example of effi- ciency being lost due to the market not protecting the individual or others is in the efficient allocation of education. Education on, say, health issues, if left to the market, would mean that only those with the ability to pay for that education would benefit, and therefore leave others more vulnerable to health risks and, as an aggregate, be detrimental to the labour resource at large. Compulsory education up to a certain age, such as a legal obligation of 16 years in the UK, means that a pater- nal approach to children means that a more equitable life chance is available for those children who may have to enter the workforce earlier if not protected.

A final social goal that is not met by the efficient allocation of resources by the market is in providing obligations. Citizens of a nation will have various social obli- gations to fulfil. These obligations may be more formal in relation to the state or less formal in relation to the community and family to which a person is affiliated. As goods and services bought and sold on the market are money-backed, the market will not necessarily efficiently provide obligations that can be easily traded. Formal state-based obligations that meet the needs of wider society are those such as voting rights. In many urban areas that operate politically under a democratic system, each person has a vote that is individual to them and cannot be bought or sold on the market. This vote will have significant influence on how the economy is handled by those elected to be in power both nationally and within the locality in which the voted member has been elected. Less-formal obligations that cannot be efficiently bought and sold on the market are those formed in affinity to a person's community or family (biological or non-biological). For instance, volun- tary work carried out in a community group (e.g. a voluntary elderly care centre) may mean that the quantity and price of care provided does not allocate an efficient provision of 'money-backed' community services (e.g. elderly care).

Familial obligations that distort efficiency in the market, and thus deem it to have an element of failure, is in, say, childcare where a parent can, in the short term, delegate responsibility for a sibling or grandparent to look after the parent's child. These obligations would mean that no price is attached to care that the market could provide, and therefore distort the fully efficient market price.

Summary

1   Markets in reality are not entirely 'free' in that the agreed price and quantity exchanged at will be influenced in some by market imperfections and from authorities that can guide and control the force of an 'invisible hand'.

2   A primary objective for producers is to sell goods as profit maximisers, where the revenue income is higher than the costs paid out over a designated time-frame. Government interest in the market may depend on the elected party in power at the time, in whether they are in favour of supporting a freer market approach to the economy – and thus influencing the urban areas contained within its national boundaries.

3   The reality is far from this simple duality of market contexts, although it enables a good starting point to realise differing forces that influence market activity to either encourage or rectify market failure and externalities. A PPP (public–private partnership) is difficult to clearly segment into clear ownership segments as the contract may shift ownership as the project develops.

4   To reach a maximised level of profit, an organisation will usually adjust influential factors such as production costs, sale prices and output. There are two main profit-maximisation methods used, and they are marginal cost–marginal revenue method and total cost–total revenue method.

5   In economics, the costs internal to the firm and the costs external to wider society and the environment are of importance, particularly as transactions will have a cost imposed on a third party not involved in the internal market transaction. Both the internal cost and external cost are referred to as the full cost in producing a good or service. If the external pollution costs of radiation are not within the market price, some other party in society will be inefficiently paying for costs from which they have not derived any utility or satisfaction.

6   Externalities can be more broadly defined as the spill-over or third-party effects arising from the production and/or consumption of goods and services in the marketplace. Transactions within the market will generate costs and benefits (positive and/or negative externalities) that are external to those producing and consuming goods and services.

7   'Market failure' is the key concept for discussion here, and is where the market fails to achieve an efficient allocation of resources. This inefficient allocation of resources certainly applies to environmental resources in their use, production and consumption within urban areas. Five reasons why markets are deemed to fail in efficiency are: (1) monopolies; (2) externalities; (3) the use of common and public goods; (4) weak property rights; and (5) asymmetric information.

8   The market also fails to achieve other socioeconomic goals. These socio-
    economic goals are particularly pronounced in urban areas where a
    more-dense population interact both socially and as isolated individuals. Three
    key goals are stressed and these involve distribution, protection and obligations.

Chapter 8

Cost-benefit analysis and discounting

Manchester, UK

The previous chapter has demonstrated that there are external costs and benefits that need to be internalised to ensure that all economic resources are considered. Externalities, such as the costs of pollution, as an obvious and relevant example, will need to be measured and valued if they are to be accounted for as full economic costs. Environmental pollution will be detrimental to another (third) party's resource inputs, and therefore has significance to the scarce allocation of resources to meet infinite wants in society. This chapter will explore a cost-benefit analysis (CBA) approach to measuring and valuing economic costs and benefits that will apply to the social and environmental development of urban spaces. Cost-benefit analysis (CBA) will be demonstrated whilst incorporating an element of discounting of money, to account for changes in monetary value over time. Measuring and valuing environmental and social phenomena that are produced, consumed and/or used in urban areas is problematic but wholly necessary if their costs (and benefits) are to be met by society in their entirety. Hence, an overview of the issues, relevance and pragmatics of valuing the environment is shown.

a. Cost-benefit analysis and discounting

Due to market failure and the market being unable to fully internalise externalities other techniques to value beyond the market are appropriate for environmental resources that are produced and used in urban areas. A well-used technique in practice that can capture social and environmental goods and services is through cost-benefit analysis (CBA). What is particularly useful in the CBA approach is that the net cost or benefit of a project can be calculated over different time periods, plus the changing value of money over this time period can be incorporated into the overall cost and benefit of a project. The application of a project is often given to CBA approaches, as it is often the larger-scale project that uncovers and brings into focus the wider social and environmental impacts that may occur.

b. What is cost-benefit analysis?

Cost-benefit analysis (CBA) makes operational the very simple, and rational, idea that decisions should be based on some weighing up of the advantages and disadvantages of an action (Pearce et al., 1990). CBA is seen as a useful way to apply monetary values when comparing like with like of both costs and benefits. CBA allows for analysis to go beyond the idea of an individual balancing of costs and benefits to a societal balancing of costs and benefits. In returning to the principles of economics and the definition of terms in Chapter 1, costs and benefits are defined according to the satisfaction of wants or preferences. If something meets a want, then it is a benefit. If it detracts from a want, then it is a cost. More broadly, to combine social and environmental costs, anything is of benefit if it increases human well-being, and anything is a cost if it reduces human well-being. In order to set some parameters of what CBA can achieve, it should be noted that CBA is not the only technique to assist in decisions on measuring and valuing environment and social benefits and costs (or in combination, well-being), particularly as other approaches may be more appropriate. Furthermore, in using CBA as a technique, there will be a need for more considered judgement in deciding when to use CBA and when an alternative approach (e.g. market analysis) will be more effective and accurate.

c. Measuring and valuing environmental costs

In walking though a typical CBA, there is now a demonstration of what environmental economic costs and benefits of pollution control can be quantified and valued. Pollution control is the specific 'project' in mind when extracting the full costs and benefit that need to be considered as of significant importance. In drawing out environmental costs and benefits of pollution control in this example, the social costs and benefits of the environmental pollution process will be included.

The first stage of a CBA is to qualitatively identify the costs and benefits of a project or initiative such as pollution control by a regulatory authority. As listed in Table 8.1, several benefits can be drawn for pollution control such as reduced death rates, as fewer people are exposed to hazardous material. At a less severe level, the benefits of improved health can be seen at a personal level (such as a fall in counts of bronchitis, strokes, and respiratory and heart disease), plus, at a wider environmental scale, an improved visibility (as in the case of light pollution or smog) can benefit activity that needs light input. Physical property benefits can be identified such as a reduced damage to structures due to the pollution control. A final benefit identified (note there will be more than five) involves the reproductive fertility of soils and vegetation for agricultural produce to flourish and add more value if pollution is put under control.

Identified costs of pollution control will be the installation and the operational and maintenance costs of pollution–control equipment. For instance, an accurate water pollutant gauge that can track certain chemical release in the production of biochemicals will have to be manufactured, installed, monitored and repaired during the regulatory control period. A second cost of pollution control is the higher prices that may be passed on to the consumer rather than the producer in order to pay for the cost of control. This passing on of costs to the consumer will depend on the elasticity of supply and demand for the particular good or service, as discussed in Chapter 6. With an environmental good with a short-term inelastic

Table 8.1 Benefits and costs of pollution control

Benefits of pollution control	Costs of pollution control
1. Reduced death rates	1. Installation, operational and maintenance costs of pollution control equipment
2. Measured health improvements	2. Higher prices for consumer if cost passed on from producer
3. Improved visibility	3. Design and implementation of the regulations; Monitoring and enforcing regulations
4. Reduced damage to structures	
5. Agricultural productivity improvements	

Source: Author

demand such as gasoline, as there is no short-term substitution effect to, say, electric fuel (unless a new, equivalently functioning, electric car is available for purchase), the producer knows that it can increase its price in response to the tax as the consumer will still purchase the fuel. A third cost identified using the example of a pollution-control initiative or project is from a policy and policing perspective. If a regulation is introduced, not only the cost of equipment is needed but also the cost of administering, implementation, monitoring and enforcement. For instance, the collection of a tax will need to be audited by an authority to ensure the correct tax receipts are being collected.

Once costs and benefits have been identified in a CBA, a measure of these particular indicators is needed in order to begin to attach a value to them. This is unless the measure in the first instance is a monetary value that makes measure-to-value conversion more straightforward. As per Table 8.2, measures of pollution damage are attributed as a percentage change of gross domestic product (GDP). In doing so, this measure allows a number to be attached to the damage of pollution, and thus provide measured benefits and costs of environmental policy introduced, such as the introduction of pollution control. Note that these measures of environmental 'damage' mean that they only provide indicators of costs and not benefits. Moreover, it should be noted that these measure are 'ball park', roughly estimated numbers, meaning that accuracy of figures should be considered when making judgements based on the numbers. Looking in more detail at the costs in Table 8.2, it is seen that increases in pollution damage have been measured as a specific percentage of GDP such as a 1.1 per cent of GDP in noise pollution in 2010, or a range of percentages of GDP such as air pollution ranging from 3.0 per cent to 4.1 per cent of GDP in 2011. Comparisons of pollution damage as a percentage of GDP are also made in Table 8.2, which enables further analysis, such as the comparison of total measures, or the top-end parameters of annual damage changing from 6.8 per cent in 2010 to 8.5 per cent in 2011.

Table 8.2 Pollution damage by % change (2010 to 2011) as a proportion of GDP

	Annual damage 2010: % of GDP	*Annual damage 2011: % of GDP*
Air pollution	2.7 to 3.8	3.0 to 4.1
Water pollution	1.3 to 1.9	1.4 to 2.2
Noise nuisance	1.1	2.2
Total	= 5.1 to 6.8	= 6.6 to 8.5

Source: Author

If specific metrics (such as % of GDP) can be applied to environmental (or social) benefits and costs, attaching monetary value can be attempted. The example used in Table 8.3, demonstrates how particular pollution benefits and costs have been identified and have monetary values attached to them. For instance, using 'dummy' values for particular urban area indicators such a health, have a monetary cost attached at between $1.7 bn and $2.8 bn due to pollution levels. Specific costs are those such as $10.7 bn attributed to house-price depreciation due to pollution levels, such as those properties that are effected by smog in downtown traffic-congested areas. Total costs in the example are valued at $49.9 bn and are an aggregate of all costs identified and valued. It should be noted that not all types of environmental damage are capable of 'monetisation' because a monetary value cannot be placed on all items. As only a limited set of items or indicators can be identified, this means that the numbers produced are often undervalued estimates. Despite some difficulties in valuing fully and accurately, they do hold a great deal of practical use. The values are particularly used in policy (e.g. environmental policy), especially if a significant cost to the economy has been demonstrated (e.g. $49.9 bn in this example).

Table 8.3 Example of pollution damage as monetary costs

	2009–11 US$ billion
Air pollution	
Health (respiratory disease)	1.7–2.8
Materials damage	1.7
Agriculture	1.0
Forestry losses	1.7–1.9
Forestry recreation	1.9–2.7
Forestry – other	1.0–1.1
Disamenity (toxic release)	16.6
Water Pollution	
Freshwater fishing	1.0
Groundwater damage	3.8
Recreation	n.a.
Noise	
Workplace noise	2.0
House-price depreciation	10.7
Other	1.6
Total	**= 49.9** (higher values)

Source: Author

As well as costs from pollution being identified and valued, as an inverse analysis of the benefits of pollution control are equally possible and provided in practice.

Table 8.4 demonstrates some of these benefits in monetary value, such as air pollution figures that include items such as improved health. If less people are using hospital resources with bronchial and respiratory problems, this will be a financial saving. Also within air-pollution controls there would be benefits such as less degradation of assets, therefore forgoing 'soiling and cleaning costs', vegetation being more productive if less polluted, as well as materials and property values increasing in value with less exposure to air pollution. Note that some costs of pollution (e.g. property damage incurred) may be netted off against the benefits of pollution control (e.g. property damage saved), although the net difference of what is actually polluted and what can be contained often differ, meaning it is best practice to consider and identify both costs and benefits of items that generate monetary values if the project (e.g. a pollution control initiative) is introduced (in this instance the benefit) or if it was not introduced (in what would be the opportunity cost). Further benefits of pollution control in water resources used by urban areas are those such as increasing the leisure revenues from cleaner water (e.g. fishing, boating, swimming and hunting), plus those such as diversionary uses of water that could be the benefits of clean water irrigation. To add further analysis to benefit, the largest benefit gained from the pollution control is in the area of health with an estimated (and note more illustrative figure) $US18 bn in 2011. By adding up the individual items to produce the total benefits of a pollution control measure it can be seen (in Table 8.4) that without control pollution the damage caused would have been $39.6 billion higher than actually realised.

Table 8.4 Example of the benefits of pollution control

	2011 US$ billion
Air pollution	
Health	18
Soiling and cleaning	4.1
Vegetation	1.4
Materials	1.8
Property values	1.8
Water pollution	
Recreational fishing	2.1
Boating	1.9
Swimming	1.6
Waterfowl hunting	1.2
Non-user benefits	1.7
Commercial fishing	1.5
Diversionary uses	2.5
Total	**= 39.6**

Source: Author

d. Time and discounting

Now that the cost and benefit items have been identified and valued, the net difference needs to be considered in relation to time. The reason that time is factored into a CBA of a particular project is to account for the change in the value of money over the project period. The value of money changes over time due to fundamental principles of how people will prefer to satisfy want now rather than later. Conversely, people will also typically prefer costs later rather than now. As an example, satisfaction of wants such as buying a house will be preferred now (from a consumer point of view), although the costs of paying for the house would be preferred sometime into the future or even deferred to the point of non-payment (if the seller allows it). This difference in satisfaction now and incurred cost at a later date, means that goods and services will typically have a higher net benefit and cost in the future, and thus this change in the good's or service's future value needs to be discounted back to today's price of money. Put differently, because individuals attach less weight to a benefit or cost of a project in the future than they do to a benefit or cost now (because the transaction is important to the buyer and seller at today's monetary value), the monetary value will have to be adjusted (or discounted) to more accurately reflect the true net cost or benefit at time periods ranging from the start of the project to the end of the project.

To apply an example, the future benefit of infrastructure in year 1 may be valued at £100 million and in 20 years may be valued at £10 million, but this value in year 20, the net benefit of £10 million, is using the cost of money 20 years ago, so needs to be adjusted. The discount rate operates like the opposite to interest if money was saved rather than invested. For instance, if money (M) is put into the bank (rather than invested in infrastructure) at an interest of 10 per cent over 20 years, the value of money in 20 years would be worth $M \times 20 \times 10\%$. To counteract for an opportunity cost in savings (% interest received from the bank) forgone when investing (in infrastructure), the discount rate of money needs to be applied for each period of time.

This method of discounting can consider the 'real' net cost or benefit of goods and services over time, but it can also be used to consider the environmental impact of a project over many years, stretching over several generations (e.g. 100 years). For instance, the net benefit of a pollution control initiative can be valued into the future whilst calculating how the changing value of money will affect the net benefit at certain time periods. Note that the discounting technique is simply to generate estimated valuations into future periods, and does not necessarily reduce environmental costs in the future. For instance, the changing of discount rates can be used to alter the amount of net benefit or cost that will be incurred towards the start of the project (say in year 5) or at the end of the project (say in year 20). If there was a reduction in the discount rate from 10 per cent to 5 per cent, this would mean that the net benefit or cost is discounted more in the earlier years and would have less consideration for future net benefits or costs. As the inverse, an increase in discount rate from 5 per cent to 10 per cent would mean that net benefit or cost would be discounted less in earlier years and have more consideration for future net benefits or costs. A more in-depth example is applied shortly, but what this means is that, if higher discount rates (e.g. 10% rather than 5%) are applied to a project that has high environmental impact costs in the future, the 'real' monetary costs of these high future costs will have been considered and dampened in the model – to account for how money will be worth less in the future. To round off this initial discussion in lay terms, discounting occurs as people would prefer to have money in their hand today rather than in the future, because money will be worth less in the future. This, in turn, means that if money is worth less in the future, any project in the future that has benefits or costs needs to be adjusted to account for such devaluation of money over time.

e. The mathematics of discounting

Now that discounting has been introduced to value net benefit or cost, the mathematics of how a discount rate is applied should be briefly demonstrated. From Table 8.5, over a five-year period the aggregated net benefit has been calculated from year 1 to year 5 as valued in currency (£, $, € etc.) of −30 (a net cost), −5, 15 (a net benefit), 15, and 15. This means that as the five-year project (e.g. an urban bypass) progressed, it went from a heavy initial cost to become a sustainable net benefit from year 3 to year 5.

$$NPV = \sum_{t=0}^{T} \frac{(Benefit_t - Cost_t)}{(1 + r)^t}$$

Figure 8.1 The formula used for applying a discount rate to a net benefit or cost
Source: Author

Table 8.5 Table of aggregated values for discounting: benefits, costs and net benefits

	Year 1	Year 2	Year 3	Year 4	Year 5
Benefit	0	5	15	15	15
Cost	30	10	0	0	0
Net benefit	−30	−5	15	15	15

Source: Author

A discount rate is then calculated to the net benefits and costs for each year. As an example here, a discount rate of 10 per cent is used that is equal to a decimal of 0.1. In relation to the formula in Figure 8.1, this means that the part of the formula that uses 1 + r (where r is the rate of discounting) the value to be used in each calculation is $1 + r$, which equals $1 + 0.1$, or if added together 1.1. From the formula in Figure 8.1, t represents time and the sigma sign represents the summation of all five years of calculation. Hence, when plugging in the numbers into a long-hand equation, the calculation that should be processed is as follows:

$$= [-30 / (1.1)^1] + [-5 / (1.1)^2] + [15 / (1.1)^3] + [15 / (1.1)^4] + [15 / (1.1)^5]$$

$$= -27.3 - 4.1 + 11.3 + 10.3 + 9.3 = -0.5$$

From this calculation it is found, as to be expected, that the final net value (NV), which is the total net benefit or cost before discounting, is different to the final discounted NPV. In this example, the figures gave an NV of 10 (benefits – costs = 45 – 35 = 10), without applying a discount. Although when applying a discount, the final discounted NPV was at negative 5. What this figure shows is that a negative total project value means that the project should not go ahead, especially as all costs and benefits over time, and changes in monetary value over this time, have been considered. Despite this, considerations could be made in practice as to whether this 0.5 value is of significant material value to not go through with the project, or alternatively, different discounted rates could be applied to ensure that there is a positive net benefit – but, most importantly, justifying the discount rate is a true and fair one.

f. NPV (net present value) and discounted values for a whole project

The short five-year introductory CBA example for the purposes of mathematical understanding can now be transposed to a more practical project template. Table 8.6 is one such CBA template example that considers the: costs; benefits; net benefit or cost; and two alternative discount rates (5% and 10%) over a 15-year period. Here, as per step 1, cost and benefit items are identified such as the actual construction and maintenance costs, and the benefits generated from shorter journey times. The second step of applying monetary values is put into the CBA with, say, 70K being applied to the third year of maintenance costs. The third step of adding together all costs and benefits is then made, with the addition of another row of calculation that is the net value (NV) of total costs taken away from total benefits. The fourth step in the process is discounting, and here both a 10 per cent rate and a 5 per cent rate are applied in order to show how the variance in rate will influence the weight in calculation of the present values, and subsequently adjust the final total discounted NPV. In the template provided, it can be seen that a lower discount rate (5%) has less of an influence in reducing the NPVs (e.g. in year 1, multiplying through by 0.95 rather than 0.91 means that a multiplication closer to 1 doesn't reduce the net benefit or cost by a greater amount). The important figure at the end of the analysis is the value in the bottom right-hand corner, which is the total discounted NPV (*L–F). If this value is positive, it indicates that, according to the figures used, the particular project should go ahead. If the value is negative, the opposite is true and the project should not go ahead. As mentioned in the previous section, the discounted value can be adjusted in the spreadsheet (if using more sophisticated tools away from first principles demonstrated here for teaching purposes) in order to generate a positive discounted NPV, although justification of the value will be required for a credible analysis to be approved.

Table 8.6 CBA template for an urban bypass proposal

Year	1	2	3	4	15	Total
Costs			(£'000)			
Bypass construction cost	35K	35K				
Bypass maintenance costs			70K	70K	70K	
Country park visitors during and after construction						
Wetland destruction						
Total cost	A	B	C	D	E	F
Benefits						
Reduced accidents and costs to health services						
Shorter journey times						
Reduction in noise, vibration, air pollution and visual blight						
Increase in house prices						
Total benefit	G	H	I	J	K	L
NV (net value)	**G–A**	**H–B**	**I–C**	**J–D**	**K–E**	**L–F**
Discounted values						
*Discount rate (10%)	1	0.9091	0.8264	0.7513	0.2394	N/A
*Discount rate (5%)	1	0.9524	0.9070	0.8638	0.4810	N/A
Net Present Value (NPV)	* G–A	* H–B	* I–C	* J–D	* K–E	* L–F

Source: Author

g. A simple discounting example: cost of nuclear waste

Whilst bearing in mind these technical steps in producing a CBA, another basic scenario that will make this analysis clearer is in the costing of nuclear waste disposal that has resulted from generating power for urbanising spaces. Here, in the CBA of building a nuclear reactor, the discounting applied to the net value (NV) figures will mean that the future NPV will change in value depending on the discount rate used. If, say, a low discount rate was applied, this would mean that future costs and benefits of installing a nuclear reactor would have (on paper) higher net cost or benefit in the future. If this future value was a cost that spanned many generations – for instance if the waste-disposal costs were higher than a reactor not generating any power or jobs the discount rate applied will make a large difference as to whether the future cost is high enough to discourage a decision to build the reactor now. Figures will now demonstrate how discounting can shift costs to future generations, particularly if nuclear waste dumps from the past can produce leakage costs to future generations.

Say if the cost of leakage is £1 billion (£1,000 million) in 100 years' time, the value (NPV) of the £1 billion in monetary terms will be far less if the value of money is less. To calculate what £1 billion is worth now (NPV), to make a decision on whether the project should go ahead now, a discount rate of 8 per cent (or as a decimal 0.08) is applied. Using the formula in Figure 8.1, and 'plugging in' the numbers, the following result of £450,000 is calculated as follows:

$$(1 + r) = 1 + 0.08 = 1.08$$

$$= [£1000 \text{ M} / (1.08) = £1000 \text{ M} / 2200 = £0.45 \text{ M}$$

$$= £450,000$$

This means that the damage costs in 100 years at today's price of money is £1 billion, but at prices in 100 years' time, the value of costs is only £450,000. From this it can be seen that the application of 'time' (particularly the time value of money) is a key determining factor of analysing future costs and benefits through discounting.

h. CBA use in practice

The future net costs or benefits that change through discounting application are a paper analytical exercise to aid decision-making on whether a project should go ahead or not. In practice, more factors need to be considered beyond the number-crunching to realise what the number can represent and how they will be biased to meet different interested parties in a project. For instance, both public and private sectors use NPV (net present value) in project appraisals, although CBA itself is primarily used for project appraisal in the public sector, as it considers public goods and services as well as private ones. For the private sector the use of financial appraisal or capital budgeting is often more appropriate than a CBA that considers external costs and benefits. In the public sector, these further external costs and benefits will use values not determined directly by the market, and by alternative methods as discussed above, such as the willingness to pay (WTP) principle or the opportunity costs (forgone opportunities – the next best alternative), to decide if a project should go ahead (e.g. build a nuclear reactor) over another project (e.g. build a school).

i.   Problems with economic valuation of the environment

As mentioned, the valuation of external costs and benefits will be difficult but necessary if the full cost or benefit to society and the environment is to be realised. These difficulties in valuing the environment should be made clear in order to qualify and justify the numbers calculated and subsequent decisions made. Firstly, environmental values should not be reducible to a single one-dimensional standard that is only expressed in monetary terms. This is because a single standards such as the value of a wetland will differ in monetary calculation depending on various factors such as location, scarcity and biodiversity of habitat. The multidimensional factors may therefore deny the ability to apply economic measures to some intangible factors. For instance, there is room for debate on how and what measured value can be applied to complex external benefits such as improved quality of life, protection of endangered species and preservation of scenic or historic sites.

Secondly, uncertainty whether some items identified actually exist may make measurement and valuation meaningless. For example, the full effects of global warming and climate change are uncertain, therefore it is difficult to fully apply a cost value of the damage. To account for such uncertainties, any calculation of costs or benefits of an item (e.g. flood damage to property) could assign a worse-case value (e.g. £2 billion worse-case cost) to the uncertain outcome of current activities (e.g. climate change). Thirdly, a problem in economic valuation of the environment is that ecological interconnections may be missed if the values of components in a system are made separately. This is because, ecologically speaking, entities in the natural environment (an animal species, a valley, a river, humans etc.) should be based on the overall sustainability (entire health) of the ecosystem. As such, to value one element in the ecosystem, such as a particular species of plant, it will be worth a different value in its connection to another species, as say a food source for another species. If this connected value is larger than the individual values (e.g. for each species), this will mean that the whole (ecosystem) is greater than the sum of the parts.

A fourth issue in economically valuing externalities such as the environment in a CBA is due to economics (urban and environmental) not being an exact 'laboratory' science of physical objects. Economics is part of the social sciences, and seeks to analyse and describe the production, distribution and consumption of resources

– but not necessarily attempt to provide absolute truths as many natural science disciplines do, such as providing a unifying theory of gravity in physics.

In valuing various environmental and social phenomena, it is important to note that this valuation process creates a new set of implications and concerns. In applying a monetary value, it implies that environmental goods and services are *not* free and can be traded in the market. This commodification enables such goods and services to be taken seriously as items that have responsibility taken for them, but it also leaves these items open for exploitation if responsibility can be put onto the market rather than a person. Valuing environmental goods and services also forces rational decision-making as the number attached to items encourages behaviour to think in terms of gains/losses or costs/benefits rather than as a moral responsibility to the environment. As a final consideration of valuing environmental resources, it is the value of 'priceless' goods that are different to the technical difficulty of valuing environmental goods and services. Here it is argued that even 'priceless' goods can be valued in models such as CBA, in that their economic value, if lost, is a cost of infinite value and thus would render a project invalid and mean that it should not go ahead.

j. Relevance and practicalities of valuing the environment

Despite some of these pitfalls and considerations in valuing environmental resources used in urban areas, there is still great practical relevance of valuing the environment (and included in a CBA) to ensure full costs and benefits are incorporated in a project. The justification and relevance of such valuation are because environmental impacts matter in society and therefore need to be measured in practice. Many government departments use a computerised approach to CBA that enables a standard programme template to be adapted to local conditions in terms of the relevant parameters over different time values. To avoid like-for-like comparison of items in monetary value (e.g. a wetland in different locations will be valued differently), some departments will explicitly reject specific monetary valuation listings but have specific listed metrics for environmental impacts. This enables practical analysis in instances where monetary evaluation is infeasible but where environmental impacts clearly matter.

From a pragmatic perspective, various methods, economic instruments and valuation are necessary as there is a need to value external costs such as the environmental impact caused by pollution. This, in turn, can generate legal valuations of what should be financially compensated to the third parties who are impacted upon whilst being external to the market transactions of others. As a large-scale example of compensation, governments can seek monetary compensation from those parties responsible for natural resources that are injured or destroyed by spills (e.g. oil) and releases of hazardous waste. Further practical reasons for valuation is in the need to value the services provided by the environment, such as in the provision of utilities like clean drinking water.

In applying an example to compensation in the justification for valuation of environmental resources, the assessment of the level of compensation from pollution (e.g. oil spill or, say, urban traffic congestion) will need careful economic analysis, particularly in assessing the level of damage. Several steps can be taken in taking the point of pollution to placing a monetary value for compensation. Firstly, following pollution, an authority will need to identify categories that may be affected (e.g. fishing industry or car drivers). Secondly, once categories have been determined, there is an estimate of the physical relationship between the pollutant emissions (e.g. oil from a tanker or pollution from a car) and damage caused to the affected categories (e.g. entities on the shoreline or resident respiratory health). Thirdly, a value judgement can be placed on the response by the affected parties to avert proportion of the damage. In essence, this is a judgement as to whether steps had been taken to reduce the risk, using the oil-spill example this would be whether the hull of the tanker was sufficiently strengthened, or in the car example, whether it had the required low emmission output. Finally, once categories, physical connectivity and assessment of risk have been determined, a monetary value on the physical damages can then be applied. Such as the compensation given according to the number of fish sales forgone due to the spill, or a lack of productivity of residents due to ill health.

Summary

1   CBA is seen as a useful way to apply monetary values when comparing like with like of both costs and benefits. CBA goes beyond the idea of an *individual's* balancing of costs and benefits to *society's* balancing of costs and benefits.

2   If something meets a want, then it is a benefit; if it detracts from a want, then it is a cost. More broadly, to combine social and environmental costs, anything is of benefit if it increases human well-being, and anything is a cost if it reduces human well-being.

3   CBA is not always the correct technique to assist in decisions on measuring and valuing environmental and social benefits and costs (or, in combination, well-being). A more considered judgement is necessary in deciding when to use CBA and when an alternative approach (e.g. market analysis) will be more effective and accurate.

4   The first stage of a CBA is to qualitatively identify the costs and benefits of a project or initiative such as pollution control by a regulatory authority.

5   Once costs and benefits have been identified in a CBA, measures of these particular indicators are needed in order to begin to attach a value to them. This is unless the measure in the first instance is a monetary value that makes measure-to-value conversion unnecessary.

6   As an inverse analysis, the benefits of pollution control (rather than the costs of pollution) are equally possible and provided in practice.

7   Now that the cost and benefit items have been identified and valued, the net difference needs to be considered in relation to time. The reason that time is

factored into a CBA of a particular project is to account for the change in the value of money over the project period. Each cost and benefit can be discounted (which can be conceptualised as the inverse of an interest rate) for each year of application according to the percentage rate applied.

8   The monetary value will have to be adjusted (or discounted) to more accurately depict the true net cost or benefit at time periods ranging from the start of the project to the end of the project.

9   The net present value (NPV) is the net figure for costs minus benefits that have been discounted for each year in operation. For the building of a nuclear reactor, the discounting applied to the net value (NV) figures will mean that the future NPV will change in value depending on the discount rate used. The future generational cost in, say, 100 years can therefore be factored into the model to provide a present value that incorporates the future cost (and benefit).

10   The future net costs or benefits that change through discounting application are a paper analytical exercise to aid decision-making on whether a project should go ahead or not. In practice, more factors need to be considered beyond the number-crunching to realise what the number can represent and how they will be biased to meet different interested parties in a project.

11   For the private sector, the use of financial appraisal or capital budgeting is often more appropriate than a CBA.

12   In public sector CBA, these further external costs and benefits will use value not determined directly by the market, and use values according to alternative methods as discussed above, such as the willingness to pay (WTP) principle or the opportunity costs.

13   The valuation of external costs and benefits will be difficult but necessary if the full cost or benefit to society and the environment is to be realised. Assumptions need to be made explicit in CBA approaches as: (1) environmental values should not be reducible to a single one-dimensional standard that is only expressed in monetary terms; (2) there is uncertainty about whether some items identified actually exist, which may make measurement and valuation meaningless; (3) a problem in economic valuation of the environment is that ecological inter-connections may be missed if the values of components in a system are made separately; (4) economic valuation of externalities is not an exact 'laboratory' science of physical objects.

14   The valuation process creates a new set of implications and concerns. Environmental goods and services become commoditised, analysis becomes focused on costs and benefits rather than moral responsibilities, and items without 'price' can be carelessly ignored.

15   The justification and relevance of such valuation are because environmental (built and natural) impacts matter in society and therefore need to be measured in practice.

16   External valuations can generate a legal basis of what should be financially compensated to the third parties that are impacted on, whilst being external to the market transactions of others.

17 Several steps can be taken from taking the point of pollution to placing a monetary value for compensation. Steps include: (1) identify categories that may be affected; (2) estimate the physical relationship between the pollutant emissions and damage caused to the affected categories; (3) place a value judgement on the response by the affected parties to avert proportion of the damage; and (4) put a monetary value on the physical damages.

Chapter 9

Macroeconomic considerations

Florence, Italy

Now that the economic techniques for urban and environmental analysis such as CBA have been provided, it is important to consider the wider macroeconomic picture and place such tools in context. This chapter will reveal the context of what is referred to as the 'open economy', in which microeconomic activity such as the operating of supply and demand in an internal market for firms occurs. This open market can be analysed in the first instance by aggregating-up demand and supply in several markets within a particular spatial boundary. These macro-considerations in the open economy will also be subject to capital goods and flows that will also be discussed in this chapter. The dissolving of national boundaries in the process of global economic activity will also be explored in the section on globalisation. The effects of globalisation in relation to how urban areas use and interact with natural resources will be the main focus of these macroeconomic considerations.

a. The open economy

In addition to the more localised economic mechanisms that operate in urban areas, the wider open economy will shape how these economic mechanisms are played out in distributing resources. The open economy can be defined as the economies of all modern advanced industrial nations (and many developing nations) that are open to large volumes of foreign trade and capital movements. This means that the macroeconomic considerations are those where large aggregate values are quantified and traded internationally and include capital goods and services (i.e. those that have the ability to generate further income in the future). These aggregate values will, to a large degree, be generated from producers and consumers in urban areas that make up the national GDP figures that trade internally and externally. So an open economy, as the term suggests, is one where economies can trade openly with others under certain rules of engagement. Most nations do trade openly and there are no closed economies (e.g. Cuba and North Korea still trade internationally), although their economic systems for distribution of resources will differ due to certain political and regulatory frameworks.

More technically, in an open economy a country's spending in any given year need not equal its output of goods and services. A country can spend more money than it produces by borrowing from abroad, or it can spend less than it produces and lend the difference to foreigners (Mankiw, 2007). This demonstrates how in an open economy the capital flow of goods and services in the form of finance moves freely between nations – and often quicker than most manufactured goods and services. To provide further technical understanding, the individual components of a nation's output in an open economy are expressed as in the equation in Figure 9.1.

$$Y = C + I + G (X - M)$$

Figure 9.1 The equation components of an open economy
Source: Author

From Figure 9.1 it can be seen that the GDP (gross domestic product) for a nation in an open economy is expressed as Y and this is equal to several components. Firstly, C is the level of consumer consumption for domestic goods and services. This consumption could be, say, the aggregate number of electric goods bought within a particular nation. Secondly, I is the level of investment made by a nation in domestic goods and services, such as investment into share capital of companies within a given national boundary. G is the level of government expenditure on domestic goods and services such as the spending on public infrastructure like publicly owned roads, rail and airports.

The important components of an open economy is the difference between exports and imports and these are represented by X – M. X is the output volume of goods and services produced within one nation and sold to other nations, whereas M is the output volume of goods and services that have been produced in another nation but consumed in the nation under study. The net difference between goods and services exported to 'foreign' nations and imported in from 'foreign' nations is referred to as the balance of trade or the net export. A positive balance of payments at the end of an accounting period (usually one year) will mean that a nation is achieving a healthy balance in that it is producing and exporting more goods than it is importing and thus is economically sustainable if this pattern occurs over the longer term.

In a similar vein, the balance of payments deals with international trading in an open economy by both trade of goods and services as well as financial capital and financial transfers. When all components of the BOP accounts are included they must sum to zero with no overall surplus or deficit. If a country is importing more than it exports, its trade balance will be in deficit, but the shortfall will have to be counterbalanced in other ways. A counterbalance could be created in several ways such as by funds earned from its foreign investments, by running down central bank reserves or by receiving loans from other countries.

For urban areas, this international trading will play a large contributing factor to its success and failure. As Goodall (1972) states, exports from an urban area are certainly the major determinant of short-run fluctuations in the level of economic activity in that area. Single basic industry areas (such as car plants) are also seen to owe their long-run growth to the volume of that basic industry located there. Despite this, economic activities of urban areas are also considered important in relation to whether 'new' export activities arise in addition to the basic industry. A decline in the fortunes of the original basic industry will result in a stagnation or disappearance of ancillary economic activity – what economics textbooks refer to as a 'negative multiplier'.

b. Macroeconomic aggregate demand and supply

So an open economy includes as an aggregate the net domestic market production, which includes aggregate supply and aggregate demand provided by a combination of both the public and private sector. When looking at markets within prior chapters there has been a focus mainly on production and consumption of particular goods and services within an urban area rather than the totals for a particular nation or between nations. Thus the international trade generated by particularly productive cities (e.g. London, Tokyo, New York) contained within nations can also signficantly contribute to the macroeconomy. In macroeconomics we therefore need to open up to the whole economy nationally and globally and understand its impact on the urban areas and the built and 'natural' environment – as well as understanding how an urban area's aggregate supply and aggregate demand can contribute to the macroeconomic position of nations around the entire world.

Aggregated demand and supply is the sum of all demand and all supply of all the markets in a given spatial boundary (e.g. a nation – UK). Aggregate demand can be defined as the total amount of goods and services demanded in an economy by companies, consumers and government bodies, including foreign participants. Also, aggregate demand is also known as total spending (*Financial Times*, 2011). Another component of the macroeconomy is aggregate supply. Aggregate supply is defined as a measure of the aggregate real production of end-product services and goods that are available in an economy in a specified time period and at a range of different prices. It is therefore, in short, the total amount of goods and services that firms are willing to sell at a given price level in an economy over a period of time (e.g. one year of accounting).

c. Capital goods and flows

Understanding of capital goods and capital flows are therefore an important feature of macroeconomic thinking in urban areas that interact with environmental resources. What is meant by capital in economics can be disentangled from two other types of capital that are those in respect to 'capital' in accounting convention and capital as wealth. Capital in accounting convention could be money invested into a business or organisation in order to generate future income. For instance, financial capital investment from a proprietor or bank into a business to purchase a machine (as an asset) will generate products and, in turn, receive a return in income from sales. With regards to wealth, capital is more colloquially the money and assets seen as the sign of financial strength for an individual, organisation or nation. This financial strength can then be used to ensure any future development or investment. With respect to economics, which is of interest here, capital is one of the factors of production (i.e. not land, labour or entrepreneurship) that is used to create goods and services rather than being an entity in the process of producing goods and services. For instance, a machine producing nuts and bolts generates finished goods of nuts and bolts, whilst the machine is not physically part of the nuts or bolts produced. This economic capital can either be a capital good (or service) that is more related to tangible items, or economic capital can be a capital-flow item that is related to less tangible (or intangible) items such as financial products. These two types of economic capital are now explained in more detail in order to apply them to urban and environmental considerations.

Capital goods

Capital goods are 'real' (as distinct from 'imagined') objects owned by individuals, organisations or governments, to be used in the production of other goods or commodities. Capital goods include factories, machinery, tools, equipment and various buildings that are used to produce other products for consumption. This term also refers to any material used or consumed to manufacture other tangible goods and services. In most cases, capital goods require a substantial investment on behalf of the organisation making a product. The purchase of these goods are usually considered a capital expense. Capital goods are important to organisations because they use these items to make functional goods for the buying public or to provide consumers with a valuable service. As a result, capital goods are sometimes referred to as 'producers' goods' or 'means of production'. The economic term 'capital goods' should not be confused with the financial term 'capital', which simply refers to money. Financial capital is considered in the next section with respect to capital flows.

As capital goods are commodified with willing producers and consumers, it means that they can be traded at agreed prices and quantities. For instance, a new stock of trains connecting urban areas within a nation can be classified as economic capital goods as they generate revenue (train ticket sales) into the future and are a 'capital' factor of production (not land, labour or entrepreneurship) in that the service that is provided (moving someone from one place to another) is created by the train rather than the train itself being the entity purchased by the ticket.

Furthermore, the stock of trains as capital goods can also be traded for more consumer goods and services. Train stock could be, for instance, produced in Germany and sold to the UK at an agreed price and quantity. In doing so, this means that nations as well as firms can have a trade balance in the buying and selling of capital goods. The balance of trade (BOT) at a national level is the net difference between what tangible goods (and services) have been traded in quantity and price. Capital goods are one component of this balance of trade as the balance of trade will also include imports and exports of consumer goods, foreign aid, spending and investment. The balance of trade is the largest component of the balance of payments (BOP), with further components of the balance of payments recorded by a nation being those such as the current exchange rate and the trading of financial items. This trading of financial items is now discussed in terms of capital flows.

Capital flows

Capital flows between nations are important to the final balance of payments figure as deficit or surplus. Also, capital flows can be considered in relation to the appropriation of financial capital held by particularly economically strong urban areas. For instance, one city may have a strong capital good production (e.g. a car plant in Sunderland) whereas another city may have a strong location base for financial capital (e.g. London as a financial stronghold, with the City as a centre for trading bonds etc.). Both exporting of capital goods as cars and capital flows as financial products will, in this case, be created in different urban areas and captured at national level as separate components, adding up to the total figure in the balance of payments.

There is therefore a difference between capital flows and capital goods in urban areas and at a national level as recorded in the balance of payments. The capital flows are not imports and exports of capital goods, they are international flows of borrowing and lending, trade in assets and liabilities. For instance, the trading of shares and bonds will mean that if there is a surplus in exporting more financial products and services to 'foreign' (rather than domestic) buyers compared to what is imported, the balance of payments value for the domestic nation will move towards a more positive position. Other financial products could be the lending of money from one domestically owned bank to foreign-owned banks or to consumers (of financial services) in another country.

This capital flow raises interesting questions as to where money is attributed to at certain points in time as the global flow of money and financial products can quickly and easily move in an era of intensifying globalisation. The influence of the process of globalisation in relation to capital goods and flows are now explored with reference to urban areas, especially as lower spatial scales below the national to urban spaces may reveal different dynamics to what capital is recorded under national macroeconomic conventions such as those revealed in the balance of payments.

d. Globalisation

Contextualising macroeconomic factors in the open economy is critical when understanding elements such as aggregate demand and aggregate supply, as well as capital goods and flows. In discussing globalisation, the term can be described as the convergence of markets, economies and ways of life across the world. A broad overview definition here is that globalisation is the worldwide process of homogenising prices, products, wages, rates of interest and profits. The important point to note here is globalisation as a process (a series of actions, changes or functions) rather than as a static stage of development. So as a 'process' of development, globalisation processes will rely on three forces for development at various scales (e.g. individual, household, urban spaces, national boundaries). The three forces for development in the process of globalisation are as follows:

1   the role of human migration;
2   international trade;
3   rapid movements of capital and integration of financial markets.

First, the role of human migration in globalisation is the way in which humans are, in general, freer to move between different jobs, spaces and nations. For instance, people can more quickly move to overseas jobs or encounter quicker commutes with the expansion of air travel and the internet (for virtual and remote working). Migratory forces enable globalisation to develop as a more mobile workforce will mean a more efficient workforce and the ability to make economic gains. The second force of international trade has had an impact on globalisation in that it is now easier than ever before to trade internationally. For example, the relaxation of employment regulations and the expansion of the EU trading area has meant that a greater number of employees are able to work within different countries without having to spend time getting work permits. The force of individuals and organisations wanting to trade internally is certainly a force that has encouraged globalisation, as greater trade will undoubtedly generate greater added value and thus economic satisfaction of wants. Thirdly, the rapid movement of capital and the integration of financial markets have had a significant influence as a force for globalisation. The forces enacted by trading buyers and sellers of capital goods are to try to make such trading easier and more efficient. As such, this trading in capital goods and flows can be encouraged to flow without barriers between national frameworks. The force in this area is therefore the greater returns that can be extracted by traders in various global locations where capital has not previously had the opportunity to flourish.

As well as forces that have encouraged globalisation processes to intensify, there are other reasons or enablers that have made this intensification happen. Three key enablers for globalisation are:

- financial markets around the world becoming more integrated;
- technology and electronic trading in the value of commodities;
- money markets in one country not being independent of world financial markets.

The integration of financial markets around the world has played a significant role in enabling globalisation to intensify. The ability of, say, HSBC (Hong Kong Shanghai Banking Corporation) to provide banking and financial services around the world is a case in point. HSBC is, for instance, the second largest banking and financial services group and second largest public company. As such it is listed on the London Stock Exchange and a constituent of the FTSE 100 index. As well as being integrated into the global financial system, it has managed to extend its reach by acquiring various banks in different countries such as the US-based Midland Bank in 1991. This acquisition is one of many integrations that enabled such a financial reach for HSBC and many other global financial organisations that operate in the global financial market. To get a sense of scale in this global financial reach, HSBC has around 7,500 offices in 87 countries and territories across Africa, Asia, Europe, North America and South America and around 100 million customers, and in June 2010 had assets of US$2.418 trillion.

With respect to advances in technology and electronic trading, they have enabled globalisation to flourish, for instance web technologies and the internet have made possible the processing of information more efficient across the globe.

The creation of a virtual space in which people work and share information has meant that activities can take place over geography without people having to physically be in proximity. As a third enabler, money markets have become more integrated with world financial markets to enable globalisation to occur. For instance, the trading of pounds, dollars, yen etc. as more liquid assets (money that can be exchanged for goods now rather than, say, bonds that can be redeemed in a few years) plays a larger role in the financial markets as a whole if finance is hard to come by as seen in the financial crisis of 2007–08. The high demand for short-term borrowing and lending in money markets (one year or less), such as treasury bills, will mean that it has an effect on the global financial market of which money markets are a part. What is of importance to globalisation, and the impact on urban areas, is that this greater inter-connectedness of money and financial markets means that investments (in say property development for cities) can more quickly and easily be financed during economic growth over a global spread; but also, liquidity for credit in buying construction materials or property assets can quickly contract following wider global financial market falls. That in turn can quickly slow down or halt the economic development of many cities across the globe.

Summary

1   In addition to the more localised economic mechanisms that operate in urban areas, the wider 'open economy' will shape how these economic mechanisms are played out in distributing resources.
2   An open economy is one where economies can trade openly with others under certain rules of engagement. Most nations do trade openly and there are no closed economies (e.g. Cuba and North Korea still trade internationally), although their economic systems for distribution of resources will differ due to certain political and regulatory frameworks.
3   In an open economy the capital flow of goods and services in the form of finance moves freely between nations, often quicker than most manufactured goods and services.
4   A nation's output (GDP) in an open economy is a function of consumer spending plus investment plus government spending plus its net balance of foreign trade (as a positive or negative value on the balance of payments).
5   Note that in an open economy the balance of payments deals with international trading in an open economy by trading of both goods and services, as well as financial capital and financial transfers.
6   A counterbalance to trade deficit could be met by funds earned from its foreign investments, by running down central bank reserves or by receiving loans from other countries.
7   The international trade generated by, particularly, financially productive cities (e.g. London, Tokyo, New York) contained within nations can also be realised in macroeconomic aggregate demand and aggregate supply.

8   What is meant by 'capital' in economics can be disentangled from two other types of capital — those of capital in accounting convention and capital as wealth.

9   Capital goods are 'real' (as different to imagined) objects owned by individuals, organisations or governments, to be used in the production of other goods or commodities. The stock of trains as capital goods can also be traded for more consumer goods and services.

10 Capital flows between nations are important to the final balance of payments figure as a deficit or surplus. Capital flows can be considered in relation to the appropriation of financial capital held by particularly economically strong urban areas.

11 There is a difference between capital flows and capital goods in urban areas and at a national level as recorded in the balance of payments. The capital flows are not imports and exports of capital goods, they are for instance international flows of borrowing and lending.

12 Capital flows raise interesting questions as to where money is located at certain points in time, as the global flow of money and financial products can quickly and easily move in an era of intensifying globalisation.

13 Globalisation can be described as the convergence of markets, economies and ways of life across the world. A broad overview definition is that globalisation is the worldwide process of homogenising prices, products, wages, rates of interest and profits.

14 Globalisation is a process (a series of actions, changes or functions) rather than a static stage of development. So as a 'process' of development, globalisation will rely on three forces for development at various scales (e.g. individual, household, urban spaces and national boundaries).

15 The three forces for development in the process of globalisation are: (1) the role of human migration; (2) the level of international trade; and (3) the rapid movements of capital and integration of financial markets.

16 Three key enablers for globalisation are: (1) financial markets around the world that have become more integrated; (2) technology and electronic trading in the value of commodities; and (3) money markets in one country not being independent of world financial markets.

Chapter 10

Macroeconomic government objectives and policy

Photo 10 Arequipa, Peru

The forces and process of macroeconomic factors can shape how urban areas and environmental resources are allocated. In this chapter we start to consider these factors with respect to macroeconomic objectives and the ways in which they can be measured for greater precision in research and practice. Consideration of the role of interest rates and exchange rates in affecting these macroeconomic government objectives are also incorporated with respect to theory and application to urban and environmental issues. Wider economic policy that can guide these objectives are further discussed with reference to three major policy tools that are either fiscal, monetary or direct-regulatory in character.

a. Macroeconomic government objectives (and their measurement) for urban areas and environmental resources

GDP and GVA

As introduced with regards to growth in production possibility (Chapter 4) and its limits (Chapter 5), the study of growth here, as one of six macroeconomic objectives, can uncover how urban areas contribute to and are influenced by wider macroeconomic forces; plus there is an understanding of how some of these objectives, such as growth, are measured. At a national level, economic growth tends to be measured as GDP (gross domestic product), although when considering urban economic growth for cities and towns, an alternate, more specific, measure for that space may be more appropriate. Economic growth for cities and towns are often measured in terms of GVA (gross value added). GVA is a measure of the value of goods and services produced in an area, industry or sector of an economy, minus the cost of the raw materials and other inputs used to produce them. For sub-national GVA in the UK, ONS uses an income-based measure. GVA is mainly composed of the income made by employees (earnings) and the business (profits/surplus) as a result of production.

GVA is often confused with gross domestic product (GDP). The difference is that GVA doesn't include subsidies and taxes on the products and services produced, notably VAT. The reason GVA is preferred to GDP for regional statistics is that it is not possible to allocate tax sub-nationally. In summary:

GVA + taxes on products − subsidies = GDP

ONS publishes figures for both total GVA and GVA per head of the resident population. As GVA data are workplace-based, the value of production is allocated to the place where it takes place rather than where the people involved actually live. This means users must be cautious using GVA per head especially in urban areas with high levels of commuting. It also means that GVA should be seen as a measure of the economic output of an area rather than its wealth.

Regional GVA are published as part of the regional accounts in the UK, and are consistent with the national accounts. The regional accounts use NUTS

(Nomenclature of Units for Territorial Statistics) geographies for consistency with Europe. Particular interest with urban areas and GVA are those regions that have a strong city-region that can demonstrate some economic growth effects of the city in question.

If regional GVA values cannot act as a reasonable proxy for sub-regional urban analysis of economic development measures, other indicators can be used. In connection with the economy and other macroeconomic government objectives, the degree of unemployment in an urban area may be an easier measure to ascertain. As per Table 10.1, it can be seen in the UK that with respect to employment in 2011, there are five cities that all have employment rates above the national average at 70.4% – Aberdeen, Bristol, Leeds, Milton Keynes and Reading. The claimant count (those claiming some form of welfare payments – e.g. JSA (Job Seeker's Allowance)) can also demonstrate some proxy for macroeconomic success for an urban area. For the UK for 2011, the figures show these same cities having equal or lower than national average claimant counts except for Leeds, which had a higher claimant count at 4.1% despite its high employment rate, which may suggest a much more protracted inequality of income and employment.

Table 10.1 Top five growth cities in the UK 2011

City	Claimant count November 2010 (%)	Employment rate July 2009– June 2010 (%)	Residents with high-level qualifications 2009 (%)	Business stock per 10,000 of population 2009
Aberdeen	2.2	78.5	41.4	341
Bristol	2.8	74.2	33.2	313
Leeds	4.1	70.4	30.9	269
Milton Keynes	3.5	72.5	33.7	379
Reading	2.2	76.2	37.9	371
UK average	**3.5**	**70.4**	**29.9**	**333**

Source: Adapted from Centre for Cities (2011)

Stable prices (connection to interest rates)

Stable prices are a macroeconomic consideration for nations and the urban areas contained within their boundaries. For the urban areas themselves, the changes in inflation, or changes in the prices of a basket of goods, may be difficult to drill down to a sub-regional and city or town spatial scale. The typical measure for inflation at national level (particularly in the UK) is the CPI (Consumer Price Index) that measures the change in price levels for consumer goods and services purchased by households. Here, a typical 'basket' of goods and services as a representative sample is selected and the change in price is recorded over a period of time (mainly over a year). A high percentage change in prices of goods would mean that inflation is running high, and this may in turn devalue in 'real terms' other economic features of an economy. For instance, if wages remain constant (at say $30,000) they will be reducing in 'real terms' if goods and services are increasing in price – fewer goods and services can be purchased with the same amount of income, hence the real wage has dropped. Also the wealth of a nation, if inflation is running high, will mean that 'real growth' will be affected as the growth figure, as GDP, could have increased simply by a rise in the price of goods and services more broadly.

The measurement of CPI is therefore critical to what level of inflation is used and applied to other economic features. What is in the 'basket' of goods and services will determine how the overall increases in prices are generated. For instance, if food is in the measured basket of goods and the price of food is increasing rapidly, so will the measure of inflation. If energy goods are included in the basket of goods and the price of oil is increasing, the measure of inflation will therefore be rising. For CPI in the UK as an example, it calculates the average price increase as a percentage for a basket of 600 different goods and services. Around the middle of each month it collects information on prices of these commodities from 120,000 different retailing outlets. The UK changed from the RPI (Retail Price Index) to the CPI in 1997 under the New Labour government (1997–2010). Interestingly, the RPI has often had a higher percentage change value to the CPI meaning that the shift to using the CPI enabled inflation to be measured at a lower rate. CPI at the lower rate excludes some property-related goods and services, such as rental and mortgage payments, and therefore can be argued to mask some property-related changes in the economy. Inflation measures will therefore not include price rises where consumers pay high housing costs through rental increases or if they pay higher mortgage (variable rate) payments if there is an increase in interest rates. In urban areas, therefore, where there is a greater proportion of property owners and economic concerns over property price and interest rates, the inter-connection and transmission of interest rates and inflation are particularly pertinent.

Nationally, a government will be concerned with maintaining stable prices as per their principal macroeconomic objectives. If measures are demonstrating that prices (or inflation) are starting to rise, this will mean that action to curb such inflationary activity may be needed. One particular lever to control inflation is interest rates, and these are often set by the bank of the particular economy responsible for its economic interests. For instance, in the UK the New Labour government allowed the Bank of England MPC (Monetary Policy Committee) to independently (politically) set the interest rate in accordance with the inflation target set by the Chancellor of the Exchequer (the head of the Treasury). If the inflation target (e.g. 2 per cent annual rise in prises) is being exceeded or not being met to within a certain parameter (e.g. 1 per cent either side of the target), the MPC will decide whether to act by raising or lowering interest rates in order to put the 'brakes on' or 'brakes off' the price changes. There will be pressures from many individuals and groups affected by changes in interest rates. In particular, there is considerable pressure from forces interested in property that are extensive in urban areas, and from investors who may choose to invest in property (or seek returns elsewhere).

If interest rates were raised, this would mean that the cost of borrowing money, or as previously discussed the value of money becomes more expensive. Using the urban environment and property in particular as an example, the result of this increase in interest rates (e.g. 2% to 5%) would mean that there would be less investment in property and a greater incentive to save and keep money in the banks. The banking system would particularly put pressure to raise interest rates as they will be able to recoup some larger interest payments. However, from an economic growth perspective, there is pressure to keep interest rates low, as it would be in the interest of the economy to have investment in goods and services so that jobs, income and wealth are promoted. Using the example of housing, if interest rates are low, the number of mortgages taken to purchase property will increase demand and thus boost the amount of building on aggregate. Furthermore, if interest rates are low, property companies can borrow more to invest in building further developments to extend or modify urban spaces. A macroeconomic conundrum in setting interest rates may occur, though, if inflation is rising but with low growth in the economy, meaning that a rise in interest rates to curb rising prices could further stifle economic growth.

Stifled economic growth has been a particularly problematic feature of economies globally where the global recession triggered in 2007–08 has meant that investment for growth, rather than inflation, has been a significant priority, along with an additional investment problem compounded by lack of credit finance for individuals and companies. The attempt to loosen credit in the system globally has been through a series of injections of money into the economic system via the printing of money (termed more recently as 'quantitative easing'). The mechanism to make this happen is, in simple terms, by writing bonds (a form of an IOU) by the national bank to its borrowers (e.g. investment and high-street banks) in exchange for receiving cash. This bond will have to be recouped at some future stage and thus will mean that in order to recoup, the lending pressure on raising interest rates will emerge at a future date when it is hoped that economic growth has returned.

Full employment

Low or negative economic growth and low investment have serious consequences for goods and services being produced and thus will have a negative impact on employment (and *vice versa*). Full employment is therefore a macroeconomic objective of government that will enable human resources to be fully utilised, social costs to be reduced and self-esteem and dignity to be maintained in the wider society. High unemployment is particularly problematic when concentrated in space, such as in urban areas. In a developed context, the rapid global shift of the manufacturing sector overseas generated a high unemployment rate for many inner-city areas where there was a predominantly non-skilled or manual labour force in spatial proximity to the production process (e.g. the factory, mill etc.). In this instance, the unemployed are not earning wages and therefore not paying income tax. Furthermore, they are receiving financial support that has to be funded by those in work. Full employment is therefore preferred, as if there are more people employed, the government has more revenue for its public spending plans.

Discussion of these employment processes are covered in more detail in the next chapter, although here it is important to ensure that we are clear on how such changes in employment are measured in order to be conceptualised more clearly and applied in policy. One such unemployment rate is sourced by the International Labour Organisation (ILO). This particular measure is used by the UK government. This is a survey-based measure that defines the unemployed as those who are without jobs and who say that they have actively sought work in the last four weeks or are waiting to take up an appointment within the next fortnight. This rate is expressed as a total figure, or as a percentage of the workforce. Alternative unemployment measures are appropriated by those claiming unemployment benefits – introduced earlier as the 'claimant count' or 'claimant unemployment'. For instance, in the UK, there can be a monthly measurement in the number of those claiming Jobseekers' Allowance (JSA). Claimant counts such as these will be enabled at an aggregate city level (to track urban change) as different local authorities will hold data on who is claiming benefits in their particular administrative boundary.

Equilibrium in the balance of payments

Nations with urban areas that act as generators of economic growth, producing high added value to goods and services will contribute to what can be exported and what is desired by the domestic economy. It is desirable as a macroeconomic objective that the balance of payments remains in equilibrium (or is positive or negative only in the short term) with export output equalling import output. For instance, at a national level from a UK perspective, the balance of payments (also referred to as the 'balance of trade' or the 'current balance') is an account of the UK's trade with the rest of the world. This is the difference between the value of goods and services imported *into* the UK and the value of goods and services exported *from* the UK. This export trade will have been, to a large degree, produced by firms in urban areas within the UK, plus the import trade will have been consumed, to a large degree, by individuals in urban areas in the UK.

So for the balance of payments to be positive, there needs to be more spending on domestic goods than on foreign imports. Conversely, it will be a negative figure if domestic consumers are spending more on imports from abroad than domestic goods. The measure of balance is preferred to be in equilibrium over the longer term as the exchange rate will eventually be affected. Currency for the goods and services will have demands placed on it in the supply of money. For instance, a government does not want a large imbalance between the amount of domestic money going abroad to pay for imports and the amount of foreign money coming into the country in exchange for exports. Persistent imbalances will affect the exchange rate and the Central Bank will need to intervene to try to correct this. To provide further introductory detail to the balance of trade at a national level, remember here that the focus is on urban areas. There are three key components of the balance of payments – current balance (trading of goods and services), the capital account (trading of finance and money) and official financing (such as interest paid on debts). Nations with urban areas that trade in vast wealth of global finance will no doubt have a large sway on the net balance of payments figure in the capital account. This is particularly true of, say, London, New York and Tokyo.

Protecting the environment

The focus of this book is urban and environmental economics so macroeconomic government objectives (such as growth) and its measures (e.g. GDP) may be played out differently depending on the proportion of significance that an urban area has on the national economy in question. Plus, for the particular urban area in question, its own economy and economic objectives may mean that priorities in the use of environmental resources will differ. For instance, with a macroeconomic objective of protecting the environment at a national level, this will have a 'greener' effect on all urban areas (such as a climate change levy on businesses); as would a city strategy to cut pollution in its conurbation, such as by the introduction of a congestion charge for cars in central city zones. Hence it can be argued that the objective of environmental protection at the national and city levels (right down to the individuals within these conceptual boundaries) will mean that the allocation of scarce environmental resources will be reshaped in future years. Environmental protection is becoming increasingly important as governments respond to fears of global warming and climate change. Furthermore, governments (as well as citizens) are also keen to be (or seen to be) following the sustainability agenda that has gained significant ground in public awareness over the last decade.

Environmental protection is rather trickier to record, capture or even value. One way of measuring the quality of the environment is by the level of pollution. Other measured themes with regards to environmental protection are energy use, loss of green space and traffic congestion, which are topical and emotive issues. Specific environmental measures include greenhouse gas emissions, total waste arising in the nation, consumption of water resources by the industrial sector, and the proportion of natural areas in a nation. Also, there is the ecological footprint approach, which measures how much productive land and water an individual, city or country requires to produce all the resources it consumes and to absorb all the waste it generates.

Redistribution of income and wealth

A vast amount of income and wealth inequality occurs at the global scale, internationally, within nations and within urban areas. Urban areas have large concentrations of inequality that can be apparent in some neighbourhoods located in close proximity. If a government is pursuing an active macroeconomic policy of redistributing income and wealth, these inequalities in urban areas may help to be diffused. The redistribution of income and wealth is more usually considered a social or political objective. Apart from improving people's standard of living and welfare, a more equitable distribution of income tends to lead to a more harmonious society.

In terms of measurement for assessing the redistribution of income and wealth, gross incomes (i.e. before tax) can be compared with the level of income after the effects of tax and benefit payments have been taken into account. In one particular urban area (or even comparisons between urban areas), figures for pre- and post-tax incomes can be compared and contrasted. Alternatively, incomes can be recorded and measured with respect to pre- and post-benefits. As such, this approach, given the intervention of taxation taken away or the provision of benefits, can provide some indication of inequality. This therefore gives a picture of how evenly incomes are distributed and how those on low incomes are made better off. Further measures of inequality are often made using a Gini coefficient. The Gini coefficient is a measure of the inequality of a distribution, a value of 0 expressing total equality and a value of 1 maximal inequality (it is sometimes expressed as a percentage ranging from 0 to 100). The Gini coefficient is usually defined mathematically based on the Lorenz curve, which plots the proportion of the total income of the population (y axis) that is cumulatively earned by different percentages of the population (see Figure 10.1). The line at 45 degrees thus represents perfect equality of incomes. The Gini coefficient can then be thought of as the ratio of the area that lies between the line of equality and the Lorenz curve (marked 'A' in Figure 10.1) over the total area under the line of equality (marked 'A' and 'B' in Figure 10.1). Mathematically the Gini coefficient is expressed numerically via the equation $G = A/(A+B)$.

b. Exchange rates

Exchange rates in a particular nation can have a significant impact on how urban areas develop (especially as discussed in relation to inflation) and thus require more inputs in the form of resources – be they natural, human or manufactured (capital). The principal focus of this book is the allocation of environmental resources in urban space, so it is worth exploring how a change in exchange rates occur and how it transmits into affecting other economic factors, which can in turn lead to a change in environmental resource use for urban areas.

An exchange rate is, in simple terms, the rate at which the currency unit of one country may be exchanged for that of another. In application, this could be the number of units of euros (€) that could be exchanged by an entity (e.g. individual, broker, institution) for one US dollar. This ratio of one currency (e.g. €) in exchange against another currency (e.g. US$) is not static and fluctuates for a number of reasons and with various consequences. These consequences will, in turn, affect activity in relation to urban areas. One consequence is in affecting the balance of payments, which was discussed previously as the difference in payments and receipts (including capital goods and flows) that are recorded for a nation. Any

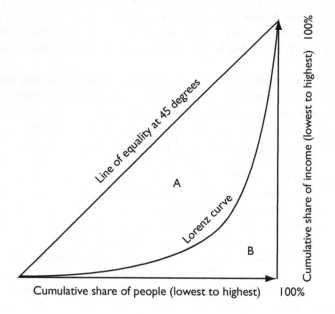

Figure 10.1 The Gini coefficient of income and wealth
Source: Author

surplus or deficit in the balance of payments will, to a large degree, be influenced by activity in the urban areas of nations where capital finance is traded and goods and services are situated.

The mechanics behind how a change in exchange rates could affect the balance of payments can be demonstrated. Firstly, exchange rates can fluctuate as for one, there will be a market for money that operates (in principle) to the economic laws of supply and demand, as described in Chapter 6. If we think in terms of supply and demand curves for the market of a particular currency such as euros in the global currency (or money) market, there will be market forces (and others such as regulatory forces) at work, which determine the value of that currency and the quantity of that currency in the global money market. If we look at Figure 10.2, which represents the market for euros in the global financial marketplace, it can be seen that the equilibrium price for that currency would be P0 and the quantity of euros available is Q0. The market for euros could (and does) shift as the demand and supply of the currency increases or decreases depending on non-price deter-minants. If we recall from the market theory on shifts in supply from non-market determinants, supply can shift due to government intervention.

Figure 10.2, shows an increasing shift in the supply of euros, and this can be generated by an expansion of the money supply through the addition of extra euros into the economy from the treasury of a nation in what is known economically as quantitative easing, or more simply, the printing of money. The printing of money today is, in essence, the addition of more zeros into the books of monetary institutions such as banks in return for a bond (i.e. an IOU) that the institutions agree to pay back at a particular rate of interest at some future designated time (e.g. 10 years). The effect we can see on Figure 10.3 from this injection of money into the system is an increase in supply from S0 to S1, an increase in quantity of the currency Q0 to Q1 and a fall in price of the currency from P0 to P1.

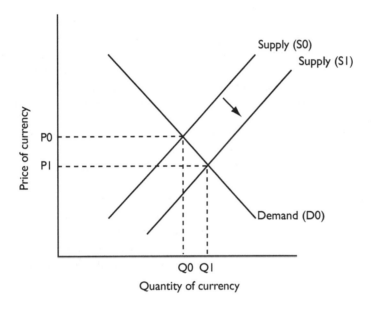

Figure 10.2 Global market for euros (€)
Source: Author

This fall in the price of the currency from a period of quantitative easing will (other things being equal), in part, be transmitted onto the value of goods and services (consumer and capital) for a nation and those urban areas within its boundaries. In the case of capital goods, if we look purely at units being sold, if more goods are being exported than are being imported, there is a surplus on the balance of payments. In Figure 10.3, 150 units are exported and 100 units are imported and this means that there is a surplus of 50 units for that particular nation. Once values of currency start to be attached to the units of this surplus, the actual value of this balance of payments surplus can be measured. As per Figure 10.3, if €1 was equal in value, or 'on parity' as more accurately described in economic parlance, it would mean that a 50-unit surplus for a country in the Eurozone would have a €50 domestic product that would be relatively equal in value to US$50. However, should the exchange rate fluctuate (e.g. from an injection of currency into the financial system) to a point where €1 is valued at $2, where €1 can buy $2 or that the € value is twice as strong as the $ value, this means that the value of the 50 units in euros relative to US dollars will change. Rather than 50 units having a value of €50 relative to $50 under the initial exchange rate at 1:1 parity, the new value of €50 would have a relative value of $100 under the new exchange rate that exchanges €1 for $2.

This stronger rate of exchange for the Eurozone country (as it can buy more dollars for €1) means that the euro is stronger in *value* relatively, even if the same amount of surplus goods were exported. This in turn means that the Eurozone country's balance of payments surplus is of higher value once the exchange rate converts all currencies to one uniform currency. If this one uniform currency was the dollar ($), this would mean that unit surplus by those trading in euros (€), once converted to dollars for uniform comparison, would be worth double the dollar ($), whereas every unit surplus in dollars ($) would not be converted and still be a dollar (at half the value relative to the euro). For the balance of payments, a country mainly trading goods and services in euros, once converted to a standard US dollar, would have a balance of payments that is stronger if an exchange rate is stronger in favour of the euro.

To bring in some application to shared urban spaces, a strong currency (in this case the euro) would mean that property in a predominantly euro-producing country would have a relatively higher value in its property assets in its cities. For example, the commercial properties within city centres would have a higher value attached to them. This may stimulate further investment into that particular city for

Capital goods

Exported goods outside UK	= 150 units
Imported goods to UK units	= 100
Net balance of payments units (+ ive, surplus)	= 50

Exchange rate relative value of units

If €1 = $1 then, 50 units = €50 domestic product relative to US $50
If €1 = $2 then, 50 units = €50 domestic product relative to US $100

Stronger €1 = stronger balance of payments for Eurozone relative to US

Figure 10.3 Units of capital goods and changes in exchange rate value
Source: Author

further urban development and expansion. Furthermore, macroeconomic factors of increasing employment and the increase in GDP (gross domestic product) or growth of output could be an advantage of such currency strength. Over the longer term, though, the strong currency may mean that its high price will discourage longer-term investment as the relative cost (compared to say a non-euro nation) of factors of production (e.g. land, labour, capital, entrepreneurship) for further development will be expensive. Conversely, this is why a 'weak' exchange rate, where the domestic currency cannot buy much foreign currency, will enable a situation where its domestic goods are cheap, and this helps it export goods and boost the economy of cities that have high production relative to consumption. A weak currency may also risk an increase in inflation as a demand-pull effect on prices may appear as more and more consumer demand is placed on 'cheaper' goods and services in the urban areas of weak currency nations.

c. Policy measures: intervention in the economy

Fiscal policy

Measuring the redistribution of income and wealth can therefore be calculated using principles of fiscal policy – taxation and government spending. The use of fiscal policy is one of three principal ways in which a government can influence how certain economic objectives are met and subsequently impact on urban areas and environmental resources used. Each nation will have a public budget in which it can make plans and allocate public funds to meet its objectives. For instance, in the UK, the government's fiscal policy is set out in the Pre-Budget Report (PBR) and in the Budget in the spring. The PBR includes a report on progress since the previous budget, an update on the state of the economy and the government's finances. It also announces proposed new tax measures. In the budget in the spring, the Chancellor of the Exchequer presents details of the government's spending plans and how it intends to finance them from taxation (and/or borrowing). Taxation represents a 'withdrawal' from the flow of income around the economy. It therefore dampens down demand for goods and services.

Specific forms of taxation can influence how urban areas are resourced and shaped. Taxation is often applied as either a charge, a percentage, a levy, or as tax relief. Examples include income tax (a tax on incomes), value added tax (VAT – a tax added on by the producer to be eventually paid by the consumer of the good or services), council tax (a tax on residence for use of local services), uniform business rate (a tax on the occupation of a non-domestic property), landfill tax (a tax on disposing of waste), congestion charge (a tax on using cars in a congested area), the climate change levy (a levy on tax on the business use of energy), and the aggregates levy (a tax on the commercial exploitation of rock, sand and gravel; this also applies to imports, to encourage recycling). As for tax relief, the government can provide a lower payment of tax to encourage certain activities. For instance, land remediation relief can encourage developers to find ways to clean up contaminated land through provision. The provision of a 150 per cent tax credit in contaminated areas of development can be provided to net off against any taxation incurred on costs.

The other side of the fiscal approach is through government spending rather than taxation. Government spending represents an 'injection' into the flow of income around the economy and therefore stimulates economic activity in the economy and urban areas. Examples of government spending are in those areas such as health, education, emergency services, infrastructure (roads, public facilities) and regeneration. Government spending can be in the form of grants and subsidies from government to encourage and support certain activities.

As such, the national government via the Treasury, or the local authority given certain central restrictions, can adjust the levels of public spending and taxation. Various tensions exist in whether a government should be high tax and/or high spend – particularly more so in a period when emphasis has focused on reduced public spending as a response to reducing the government deficit. For taxation, there would be an incentive for business to invest if they are paying lower taxes, although this would mean that there is less money going into the public 'pot' for spending or reducing the deficit. The risk of less public spending to reduce a deficit is what is referred to as 'the paradox of thrift' where less public spending will in turn reduce the amount of aggregate demand (as spending by consumers) and thus further dampen economic growth. The paradox of being thrifty by an economy is that economic growth is stifled and reduces the future output level and incomes used as a future tax base. If there is a continued approach that, for instance, promoted low tax and high government spending, this would mean that any short-fall would have to be taken from other features of the economy, such as from a balance of payments surplus.

Monetary policy

In order to manage and direct the economic resources of nations, other government intervention can help shape urban areas in the form of monetary policy, although there is limited direct influence in what governance in urban areas can control via this lever. Monetary authorities are concerned with any one (and only one) of the following levers: (1) the supply of money; (2) the price of money (interest rates); and (3) the availability of credit. In many countries the central bank is responsible for implementing monetary policy. For example, the Federal Reserve of the USA, the Bank of England in the UK and, before the euro was introduced, the Bundesbank and the Banque de France.

Principally within monetary policy, setting the interest rate (the cost of money) can have particular influence on the rate of change in prices (inflation). In the UK, as previously mentioned, the government specify is the target rate of inflation (currently 2 per cent as measured by the Consumer Prices Index), and the (independent) Monetary Policy Committee (MPC) of the Bank of England decide how to achieve this. The MPC sets the base rate (of interest) and all other financial institutions set their rates accordingly.

If interest rates are raised, it becomes more expensive to borrow money. This generally results in fewer mortgages, fewer loans being taken out, and a reduction in consumer spending. This in turn results in a fall in demand for most goods and services, which will mean suppliers are not likely to raise prices and may even reduce prices to make a sale. If interest rates are reduced, it becomes cheaper to borrow money. More people are likely to take out mortgages and to borrow for consumer spending. An increase in consumer demand will mean that suppliers will be able to sell their output, and if supply is low relative to demand, it may result in a rise in prices with regards to the supply of money, an increase in the supply of money (e.g. quantitative easing can be made and has been highlighted in the previous section on exchange rates).

Direct or regulatory policy

The third way in which intervention in the economy can be adopted, in a more command-and-control economic system approach, is via direct or regulatory policy. Note that this regulatory approach is not mentioned in all economics texts for some reason (it may be perceived as a purely legal rather than economic source of intervention). We use the term to refer to forms of direct control or direct intervention. These measures are 'objective specific' and often involve regulation or legislation. Examples of direct legislation or regulation in urban areas and in relation to building for urban development include building regulations, planning controls, business regulations and right to buy (a policy in the UK to enable purchase of your socially rented property from the local council).

In using such directives and regulations it enables the government to encourage certain activity in the economy and how the economy functions in distributing resources. For instance, regulations can enforce standards and these standards could improve environmental performance with less environmental waste. It may also improve standards of construction, safety and business behaviour. Regulations could also protect consumers such as by ensuring that consumers receive what they are informed is being sold to them, as such enhancing consumer rights. Furthermore, regulations in the built environment may also be endorsed by government to promote political objectives such as the sell-off of public assets such as infrastructure and social housing in order to encourage privatisation and perpetuate an ideology that aspires private ownership and distribution mostly via the free market.

Summary

1   Economic growth for cities and towns is often measured in terms of GVA (gross value added). GVA is a measure of the value of goods and services produced in an area, industry or sector of an economy, minus the cost of the raw materials and other inputs used to produce them.

2   GVA is mainly composed of the income made by employees (earnings) and

the business (profits/surplus) as a result of production. The difference is that GVA doesn't include subsidies and taxes on the products and services produced, notably VAT. GVA should be seen as a measure of the economic output of an area rather than its wealth.

3   The Regional Accounts use NUTS (Nomenclature of Units for Territorial Statistics) geographies for consistency with Europe. Particular interest with urban areas and GVA are those regions that have a strong city-region that can demonstrate some economic growth effects of the city in question.

4   For the urban areas themselves the changes in inflation, or changes in the prices of a basket of goods may be difficult to drill down to a sub-regional and urban spatial scale.

5   The typical measure for inflation at national level is the CPI (Consumer Prices Index) that measures the change in price levels for consumer goods and services purchased by households. What is in the 'basket' of goods and services will determine how the overall increase in prices is generated.

6   CPI in the UK at the lower rate excludes some property-related goods and services such as rental and mortgage payments, and therefore can be argued to mask any property-related changes in the economy. Inflation measures will therefore not include price rises, where consumers pay high housing costs through rental increases, or if they pay higher mortgage (variable rate) payments if there is an increase in interest rates.

7   In urban areas, therefore, where there is a greater proportion of property owners and economic concerns over property price and interest rates, the inter-connection and transmission of interest rates and inflation are particularly pertinent.

8   One particular lever to control inflation is through interest rates and this is often set by the bank of the particular economy responsible for its economic interests.

9   If interest rates are raised the cost of borrowing money, or the value of money, becomes more expensive. Using the urban environment and property in particular as an example, the result of this increase in interest rates (e.g. 2% to 5%) would mean that there would be less investment in property and a greater incentive to save and keep money in the banks.

10   Stifled economic growth has been a particularly problematic feature of economies globally where the global recession triggered in 2007–08 has meant that investment for growth rather than inflation has been a significant priority, along with an additional investment problem compounded by lack of credit finance for individuals and companies.

11   The attempt to loosen credit in the system globally has been through a series of injections of money into the economic system via the printing of money (termed more recently as 'quantitative easing'). The mechanism to make this happen is, in simple terms, by writing bonds (a form of IOU) by the national bank to its borrowers (e.g. investment and high street banks) in exchange for receiving cash.

12 Low or negative economic growth and low investment has serious consequences for goods and services being produced and thus will have a negative impact on employment (and *vice versa*).

13 Environmental protection is becoming increasingly important as governments respond to concern over global warming and climate change.

14 Measuring environmental protection is rather trickier to record, capture or even value. One way of measuring the quality of the environment is via the level of pollution or tax receipts for landfill.

15 If a government is pursuing an active macroeconomic policy of redistributing income and wealth, inequalities in urban areas may help to be diffused. The redistribution of income and wealth is more usually considered a social or political objective.

16 In terms of measurement for assessing the redistribution of income and wealth, gross incomes (i.e. before tax) can be compared with the level of income after the effects of tax and benefit payments have been taken into account. Further measures of inequality are often made using a Gini coefficient.

17 Exchange rates can fluctuate as, for one, there will be a market for money that operates (in principle) to the economic laws of supply and demand. A fall in price of the currency from a period of quantitative easing will (other things being equal), in part, be transmitted onto the value of goods and services.

18 A strong currency (e.g. in this case the euro (€) would mean that property in a predominantly Eurozone producing country would have a relatively higher value in its property assets in its cities. For example, the commercial properties within city centres would have a higher value attached to them. This may stimulate further investment into that particular city for further urban development and expansion.

19 A weak currency may also risk an increase in inflation as a demand-pull effect on prices may appear as more and more consumer demand is placed on 'cheaper' goods and services in the urban areas of a weak currency nation.

20 The use of fiscal policy is one of three principal ways in which a government can influence how certain economic objectives are met and subsequently impact on urban areas and environmental resources used.

21 Specific forms of taxation can influence how urban areas are resourced and shaped – taxation is often applied as either a charge, a percentage, a levy, or as tax relief. Government spending represents an 'injection' into the flow of income around the economy and therefore stimulates economic activity in the economy and urban areas.

22 The risk of less public spending to reduce a deficit is what is referred to as 'the paradox of thrift' where less public spending will in turn reduce the amount of aggregate demand (as spending by consumers) and thus further dampen economic growth.

23 Monetary authorities are concerned with any one (and only one) of the following levers: (1) the supply of money; (2) the price of money (interest

rates); and (3) the availability of credit. Principally, monetary policy is concerned with setting interest rates to achieve price stability.

24 The third way in which intervention in the economy can be adopted, in a more command–and–control economic system approach, is via direct or regulatory policy.

Chapter 11

Shared urban space

Themes and application

Baltimore, USA

This chapter will begin to apply key themes and examples related to urban spaces in order to illuminate the concepts and theory relating to the distribution of environmental resources and urban issues. Of particular focus is the conceptualisation of city modelling and agglomeration, as well as the environmental implications during the process of urban growth and sprawl. Economic considerations of key themes in urban areas are then explored in detail with attention to housing, neighbourhoods, employment and services (education, health and crime). Transport and other connected infrastructure are introduced, as the connectivity of urban spaces will affect the distribution of resources. As a final theme, issues of equality in urban areas are given important attention, as resource allocation have 'affects' and 'effects' that can be disproportional.

a. City modelling and agglomerations, urban growth and sprawl

It is important to conceptualise the spatial patterning and change over time when understanding shared urban spaces. In demonstrating such generalised patterns in many urban areas, several key general spatial models have been conceptualised in theory and are now explained.

Concentric zones (Burgess model)

The first significant land-use patterns that have developed in urban areas, and which in turn explain the distribution of social groups in an urban area, is represented spatially by Burgess's concentric zone model (see McKenzie *et al.*, 1967). As per Figure 11.1 the 'city' model is depicted as a series of five concentric rings rippling out from a central business district (CBD). The CBD is comprised of office buildings, hotels, museums, retails buildings etc. The first zone (labelled 1) is comprised of wholesale buildings where warehouse and storage facilities are located to provide accessible supplies for the businesses in the CBD. Out from this, in zone 2, is an area that is more transitional with a mix of low-quality (even slum) housing and a variety of mixed-use businesses. The third zone is comprised of mainly low- and middle-

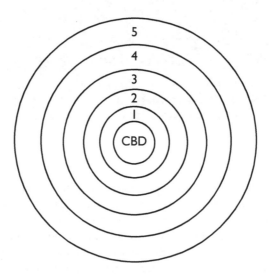

Figure 11.1 The concentric zone model
Source: Adapted from McKenzie *et al.* (1967)

income housing that would primarily house industrial workers. The fourth zone would be where upper-income, single-family residents are housed, and zone five is where high-income suburban commuters would live. Further development of this zoning model would be a growth of the CBD as the central seat of activity, and as it grows outward there is pressure on each successive zone to expand further in proportion to the CBD growth. It is described by Brown (1974) that there is a succession of land uses over time. Housing will 'filter down' from higher- to lower-income families. In the case of an eventual decline in growth (particularly in population), the outer zones would remain intact while the second transitional zone would recede into wholesaling and CBD zones. Furthermore, this would see a contraction of the CBD and an extension of the slum or blighted area of the city.

Economic explanations can be drawn from this concentric zone model in rela-tion to rent price for land and transportation costs. With respect to land-rent costs, it would, in the first instance (as shown in Figure 11.2), that businesses and people with high incomes would tend to be closer to the centre. For instance, CBD businesses at the centre would be paying high rents of R1 (Figure 11.2). However, high-income residents in zone 5 are paying low rents of R5 and choose not to outbid lower-income residents located in perceivably higher-rent land within zones 2 and 3. Explanation for this logic is given by Brown, firstly in terms of spec-ulation as, in the transitional zone (zone 2), the discounted present value (see Chapter 8 on discounting) or price of land is very high, since speculators hold the land in its present use waiting for huge capital gains when the land is transferred (mainly in agreement by the planning system) to a higher use.

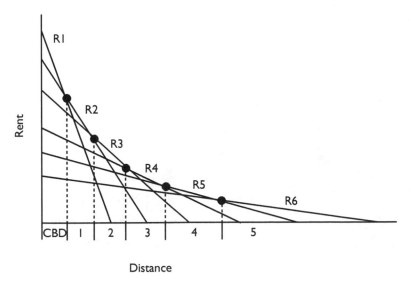

Figure 11.2 Ceiling rents and concentric zones
Source: Adapted from Haig (1926)

Low-income housing is therefore available as speculators keep up the structures on the land at a low cost and low quality. The subdivision of old houses into many apartments during the development process helps speculators increase the rent per unit of land whilst each individual apartment is kept low enough for low-income families to afford (albeit smaller and higher-density units) – this further keeps speculators with good returns during the waiting period. The further explanation on low-income housing being located near high value CBD land is due to transportation. It can be argued that residents have a preference for larger plots of land rather than distance to travel to the centre. As the outer regions are cheaper, people will trade greater land in exchange for greater commuter costs.

Radial sectors (Hoyt model)

The radial sector model of spatial patterns in urban development was developed by Homer Hoyt (1939). To further develop an understanding of Chicago (again, many of these early spatial-economic models were US-centric), the radial sector model described urban land patterns as characterised as sectors (or wedges) running out like ribbons from the CBD at the centre – based on transportation links following railroads, highways and other transport arteries (Figure 11.3). Higher levels of access enable higher land values, meaning that many commercial functions would remain in the CBD but manufacturing functions would develop in a wedge surrounding transportation routes. Housing would grow in wedge-shaped patterns with a sector of low-income housing bordering manufacturing/industrial sectors (traffic, noise, and pollution make these areas the least desirable) while sectors of middle- and high-income households were located furthest away from these residential functions.

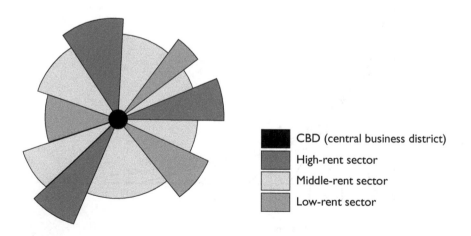

CBD (central business district)
High-rent sector
Middle-rent sector
Low-rent sector

Figure 11.3 The radial sector model of urban land use
Source: Adapted from Hoyt (1939)

Multiple nuclei model

The third well-renowned, spatially focused concept of urban land-use patterns is the multiple nuclei model as developed by Harris and Ullman (1945). In terms of the layout of the city, the model demonstrates that in addition to the CBD, similar industries with common land-use and financial requirements are established near each other and thus influence their immediate neighbourhood. The result is a patchwork of nuclei that further generate various types of land use around the nuclei (Figure 11.4). The theory has an emphasis on transport in terms of urban growth and that people have greater movement due to increased car ownership in particular. This increase of movement allows for the specialisation of regional centres that, for instance, focus on heavy industry or business parks. It is argued (Brown, 1974) that the multiple nuclei model is more realistic than the concentric zone and radial sector models, although many of the characteristics of the latter two models are included in the multiple nuclei hypothesis.

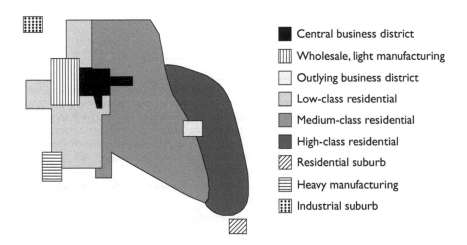

■ Central business district

▥ Wholesale, light manufacturing

☐ Outlying business district

▦ Low-class residential

▦ Medium-class residential

▦ High-class residential

▨ Residential suburb

▤ Heavy manufacturing

▦ Industrial suburb

Figure 11.4 Multiple nuclei model
Source: Adapted from Harris and Ullman (1945)

Critique of this type of modelling is in its generalisation, in that it cannot be directly applied to any one city. Interestingly, the concentric zone model (and sector model) was conceptualised when thinking of Chicago, a city that is on the coast and would not therefore not have a complete ring-growth pattern. Further difficulties of the model are that heavy industry is not sophisticatedly considered in the model. For all three models, planning constraints such as zoning are not clearly accounted for. The models do hold, though, as a good way to start thinking of how large cities in particular have developed; and there are more intricate iterations when modelling urban patterns.

Agglomeration economies

Multiple nuclei models begin to incorporate ideas of agglomeration. For instance, the gathering urban development involving various land uses around a particular nucleus is what could be described as an agglomeration. For instance, economic development around a nuclei of two or more universities in close proximity could be regarded as an agglomeration – forming a centre or hub for economic potential greater than the sum of its parts. Agglomeration economies, originally conceptualised as 'local industries' (Marshall, 1920) are principally that cost reductions can occur because economic activities are located in one place (McDonald and McMillen, 2008), and thus form an economy of scale due to spatial concentration of particular industry. As an introduction to the different types of agglomeration economies that can be conceptualised, Rosenthal and Strange (2003) have four key classifications as highlighted in Table 11.1.

Table 11.1 Classifications of agglomeration economies

Agglomeration classification	Descriptor
Industry	A variation from localisation economies (a smaller local group of industries within a single industry – e.g. pharmaceutical manufacture) up to urbanisation economies (size of the urban area includes a particular economic sector – e.g. environmental technology) where all points in between are possible sources of agglomeration effects.
Geographic	Where the effect is attenuated by distance between two establishments.
Temporal	Where the effect takes place over time.
Organisation of industry	In which the level of competitiveness has a positive effect on productivity.

Source: Adapted from Rosenthal and Strange (2003)

Urban sprawl

A final key conceptual pattern and dynamic when introducing the development of urban spaces is the notion of urban sprawl that radiates away from a central business district core. This sprawl can be somewhat extreme if drivers such as population (not always a key reason) and rising income continue to rise (enabling desires to drift further into the suburbs), as well if regulation and directives in policy, such as via the planning system, curb and shape such horizontal urban expansion. The extent of urban sprawl for some major cities such as Los Angeles in north America, and Mexico City in South America has been extreme. Low natural barriers such as mountains and great lakes, in addition to planning that promotes low-density property development, have enabled such sprawl. Density is one key measure that can demonstrate such sprawl and is calculated as the percentage of a metro area's population that lives in urbanised areas.

Issues of urban sprawl certainly relate to environmental resource considerations such as pollution, as well as raising urban social issues such as the development of homogenous enclaves of culture (e.g. the monoculture white middle-class suburb produced through white flight away from the city centre). Environmental resource problems encountered in the process of urban sprawl are those such as the reliance on car use for transportation. Other environmental impacts where pollution leading to poor health, increasing reliance on fossil fuels – leading to increased CO_2 emissions as a contributor to global warming and climate change, the increased risk of car-related deaths, and a fall in the quality and quantity of land and water. Social problems associated with urban sprawl are factors such as a reduction in social capital (an amount of connectivity to other people to enable support), as social interaction with neighbours, increased obesity due to less walking, increased transport costs affecting lower socioeconomic groups, increased costs in infrastructure as more utilities (electricity, water, gas, sewage, communications) are needed within the sprawling area, and public transport costs increasing per unit as more people turn to private car use. Further social problems of urban sprawl are those such as the social homogenisation of suburbia containing residents of the same race, background and socioeconomic status (described as 'white flight' above). Furthermore, if there is a spread of wealthier groups into previously poor neighbourhoods (or low-income enclaves as the urban area engulfs some settlements), there may be problems of gentrification – where the incumbent residents begin to be priced out (especially in rent) of the area.

b. Housing

Urban areas are home to thousands, if not millions, of economically active residents. As such, satisfying the housing demands and needs through the often scarce availability of land and property can generate resource allocation problems that are both environmental and social. Demand for housing is what economists refer to as 'effective demand', that is the money-backed desires to consume (or invest in) housing. This is different to the 'need' for housing, especially as housing is one of humankind's most basic material needs. The analysis of housing therefore should consider its multifaceted nature as a commodity that is a social good (to ensure needs are satisfied), a consumption good (for people to gain satisfaction from buying a house and living in it) and an investment good (to buy in return for some form of financial return – such as rental income or asset growth).

The first key introductory area of housing in urban areas with respect to economics and the environment is considering that the supply and demand is at constant disequilibrium in market terms. The features of supply and demand, and reasons for them not 'efficiently' meeting, are important for the discussion of the allocation of housing. The different types of disequilibrium can, firstly, frame the different types of allocation issues (Balchin et al., 2000). These include: (a) a static disequilibrium, where, at a snapshot in time for an economy, it can be demonstrated that, say, the supply of housing is not enough for the demand for housing by households (e.g. a housing deficit at the national scale in the UK); (b) a dynamic disequilibrium, where this supply-and-demand deficit (or surplus) is changing over time (such as a burgeoning deficit for housing increasing over the last decade); (c) a spatial disequilibrium, where there are deficits (or surpluses) of housing in particular regions or urban areas; and (d) qualitative equilibrium, where an acceptable standard or quality of living, seen as acceptable by society, is not attained – such as having acceptable size, condition and amenities such as heating, internal WC and fresh water.

The features and determinants that maintain disequilibrium can also be discussed in terms of how the housing market cannot efficiently distribute resources. This market inefficiency is known as 'market failure' (Chapter 7) and has several key reasons on either the supply side, the demand side, specific to the particular market as a whole, or due to features external to the market (as externalities).

Supply-side considerations

On the supply side there are features that generate many different sub-markets or sub-types. By supply type, there are differences such as detached, semi-detached, flats, bungalow, terrace etc. The conditions of occupancy, or tenure, will also differentiate their sub-type or sub-category – such as private or social rented, owner-occupied, leasehold, buy-to-let etc. Sub-markets can be location-specific in the supply of housing, such as where within a city houses are located, i.e. in the city centre or in the suburbs. Age is a further sub-category that makes the supply even more heterogeneous, with some houses being old (say Victorian or Georgian period) or being supplied in the 1970s, or being of 'new-build' or even 'off-plan'. Houses are further subdivided according to their size or residential capacity, such as being a bedsit, one-bedroom, two-bedroom, and so on. Division between public- and private-sector housing can also be differentiated, or be a combination, with a registered social landlord (RSL), such as housing associations that operate to serve public need whilst being part-financed by private funds, collecting rents that are in part determined by a percentage of market rate (e.g. 80 per cent market rents).

A significant feature particular to the housing market, and one that hinders the efficiency of the market, is the time-lag effect that makes it difficult to physically supply housing in relation to demand and price changes in the short term. For instance, a house or housing estate may take months or even years to complete and thus restrict how quick an equilibrium is met. The demolition of housing supply in some instances may also not instantly take place as properties are left abandoned for several years. This is referred to as 'inelastic supply' of housing and has been covered in more detail in Chapter 6 – the key point being that housing is built to last (Garnett and Perry, 2005). Another key supply feature in housing is that the second-hand housing market is sensitive and responsive to the new building of stock. For instance, it is noted that only 2 per cent of the housing stock is replaced each year (in the UK) and therefore any stickiness in the response of the new hous- ing sector is reflected and magnified in the second-hand market (Button, 1976). This phenomenon of new-build influence on the second-hand market is further exacerbated with only approximately 10 per cent of housing stock (particularly in the UK) being up for sale at any one time, with households sometimes choosing to hold on to their property in time of depressed housing markets. Lack of compe- tition in the supply of housing is also recognised by many (Bramley et al., 2004) as the residential property sector is dominated by a few large housing development companies. This monopoly by a few bulk-build developers means that the price can be controlled to some extent in holding prices artificially high, as the lack of competition will not force prices down.

Demand-side considerations

Several features in the demand for housing are apparent that make it unique and on a mass scale for development in urban areas. For most consumers the purchase of a house is the single biggest outlay. The fact that the purchase will generally be of one house means that it is different to many other consumer goods. The high cost of housing means that investment is often carried out via a 25-year, or more, mortgage (loan), and thus the influence of investment and finance factors cause demand to be connected to the availability of households to obtain credit and at a favourable interest rate. So in essence, one peculiarity of housing demand is the high cost of capital to purchase the 'product'. Demand for housing is related to the functions of population and income, meaning that housing's relationship with income distribution will impact on its demand. Employment considerations in urban areas are covered in the next section, and it must be remembered that imbal- ances in the market for housing, to a large degree, mirror spatial employment imbalances. For instance, the over-supply of property, say, in industrially restructur- ing cities (e.g. Liverpool, Pittsburgh) relates to the rapid fall in population and employment opportunities, and the under-supply of properties in other cities (e.g. Cambridge, San Francisco) are, in part, due to population increase and having access to more employment opportunities.

Wider housing market considerations

Despite the heterogeneity of housing supply and many characteristics of demand, the substitutability between the various categories of housing means that price changes in one market may have considerable impact on prices in other housing markets (Button, 1976). For instance, the rise in prices of the London housing market at city level is widely argued to have an impact on other city housing market prices in the UK. Imperfect knowledge (referred to in some texts as 'asymmetric information') of the different markets and sub-markets are another issue that makes the housing market more susceptible to failure. It is argued that information is a necessary lubricant to keep markets working smoothly (LeGrand *et al.*, 2008), and therefore if they are not, then a perfect equilibrium is rarely met. For instance, one person's knowledge of a particular housing market may be far more detailed than that of another, and as a result creates a situation where demand or supply operate differently according to each individual's knowledge – e.g. of where a new 'up-and-coming' housing market is due to appear.

Both the supply and demand for environmentally sustainable housing has also been a significant driver in the market. The development of zero-carbon homes has been one particular feature that has encouraged both the supply and demand of such properties. This has been, in part, encouraged in the UK by ambitious government plans for all new homes to be zero-carbon from 2016. Interestingly, consensus on a definition of zero carbon is not forthcoming, and the targets set are being relaxed in part due to the high cost of building such homes, particularly post-recession 2007–08. The zero-carbon measure will be a home that meets the code for sustainable homes (CLG, 2006) at level 6. Under the code, homes are given star ratings at six levels, with level 1 requiring thermal efficiency and level 6 being 'zero carbon'. As a new definition is being drawn up, the new proposals for zero-carbon home qualifications have flexibility in mind, in addition to reliance on in-built energy efficiency and on-site renewable energy generation. A range of additional, mostly off-site, solutions (called allowable solutions) would be made available to developers as ways to meet the zero-carbon standard.

External market considerations

Factors that are external to the market (but could in some way be valued – e.g. by CBA) will be the reason for markets not being efficient in their allocation, and thus leaving housing either under- or over-allocated in the short to medium term. These externalities have been discussed in more detail in Chapter 7, but with respect to housing some elements are worth mentioning here. Remember that externalities are more broadly seen where the actions of consumers and producers have an impact, positively or negatively, on a third party or on society more generally (King, 2008). A positive externality for housing could be, say, the neighborhood's spill-over improvements from residents keeping their homes and gardens maintained; or a negative externality could be, say, the increased public infrastructure costs or pollution generated from a new housing development. Again, these are costs and benefits borne by those third parties external to the market and are potentially measured and dealt with via a cost-benefit analysis (CBA) (see Chapter 8).

Issues relating to housing in urban areas supporting the needs of individuals and households are that the provision of housing may not be allocated according to market price and quantity logic. Housing as a social good means that it can be viewed as a merit and public good beyond the market. As for housing as a public good, it means that public (or social) housing as a tenure can be used by more than one person at a time (e.g. a public housing estate funded by public funds for many users in need) and is difficult to market to individuals based on price (i.e. it is a universal allocation system based more on need than discrimination via an ability to pay). Public provision of social housing is therefore allocated, as society deems that people to be housed based on need is socially desirable, and thus a non-market solution to the problem is provided through government via general taxation (Sawyer, 2005).

A further consideration of housing as a social good (in addition to being a consumer and investment good) beyond what the market can fail to provide efficiently is that housing is also a merit good – one that is regarded as too important to be left to the market to provide. If society deems shelter as too important to be left to the market to provide, it will mean that the allocation of housing will have to be made by forces outside of the free market (i.e. via some form of intervention). Affordable housing of a decent standard is, arguably, not attained if left completely to the free market, and therefore intervention in the market will have to occur. If residents are to be housed they will need to be able to afford the rent, and thus if society deems everyone being housed at a decent standard as socially desirable, and of merit, intervention in the market is necessary – and as such, housing is a sector that is highly regulated. Policy considerations as intervention are covered in Chapter 13 and will consider various options in urban development (including housing) that may include: (1) subsidy and taxation; (2) regulation; (3) direct provision; or (4) altering behaviours via exposure to information, publicity and rhetoric.

Housing and neighbourhood characteristics – hedonic modelling

Housing price and value is often linked to the neighbourhood characteristics in which the house is situated. A whole range of variables could therefore account for determining the price. What is used in economics to test and research the value and price of property is via hedonic modelling. As an introduction, these models calculate (via a statistical regression technique) how the house price as a dependent variable increases or decreases depending on the changes to the other variables that are characteristics of the house and neighbourhood. Outside of the science in research, the household will not obviously take all of these considerations into account. More realistically, it is realised (McDonald and McMillen, 2008) that the household specifies a desired set of characteristics in the housing unit itself in a price range, and the real-estate agent identifies units that meet these criteria. If a subject house meets the basic criteria, the household will take a look at the neighbourhood characteristics that are regarded as most important (e.g. school quality, access to shops etc.). As a quick, non-exhaustive list of potential housing units and neighbourhood characteristics, Table 11.2 is a good initial point of reference.

Table 11.2 Housing unit and neighbourhood characteristics

1. Quality of local school (positive effect)
2. Crime rate in area (negative effect)
3. Property tax rate (negative effect)
4. Neighbourhood income, size of houses in neighbourhood etc. (positive effect)
5. Air pollution (negative effect)
6. Airport noise (negative effect)
7. Proximity to contaminated areas (negative effect)
8. Proximity of nuclear power plant (negative effect)
9. Proximity to park (positive effect)
10. Industrial noise (negative effect)
11. Heavy traffic on street (negative effect)
12. Location in floodplain (negative effect)
13. Distance to employment (negative effect)
14. Distance to shopping (negative effect)
15. Distance to airport (negative effect)
16. Rating of the quality of houses next door and the block (positive effect)
17. Within walking distance of public transport or commuter rail station (positive effect)
18. Adjacent to rail line, highway or transit line (negative effect)
19. Proximity to highway interchange (positive effect)
20. Proximity to a church (negative effect)
22. Previous price increase in neighbourhood, to capture expectation of future price increases (positive effect)

Source: Adapted from McDonald and McMillen (2008)

c. Labour and employment

The connectivity of residents to their place of work will be significant to how urban spaces are developed. Spatial models of growth earlier in this chapter demonstrated both the importance of transport and the development of the CBD (in concentric-zone and sector models) and hubs of economic activity (in the multiple nuclei model) in urban spaces. For urban areas at a global level and in relation to globalisation, the impact of new information technologies and the speed of globalisation have quickened and their reach has broadened. These technologies are reinforcing the importance of knowledge and information in economic transformation, while reducing the relative importance of traditional manufacturing and industrial development based on raw materials. In urban areas, this has manifested itself in the growth of the service sector in both absolute and relative terms (UNEP, 2002). Technology has increased the already dominant economic role and importance of urban areas, not just those in the more developed economies but globally

(World Bank, 2000), indicating the growing importance of cities in the global economy.

The impact of changing global processes has been varied in different cities depending on their previous development and the specific mix of employment sector and industry types – manufacturing, commercial, retail, office, finance etc. The fortunes of 'developed world' cities depended largely on their ability to respond to the challenges of globalisation. For some heavy manufacturing-dominated industry, cities in the 'rustbelt', such as the motor industry in Detroit, the decline and subsequent population loss has been stark. Other cities in the position to develop high-technology goods and services, such as those cities in the 'sunbelt' of the US (e.g. Los Angeles, San Diego), have been more fortunate in experiencing a population increase with rising incomes and continued added value to goods and services. For cities in the developed world, these opportunities of economic growth have been enabled and have been witnessed by a rising urban middle class in the cities of, say, Mumbai and Beijing. However, the losers in this process through deregulation and increased foreign direct investment (FDI) have created a more unstable and uncertain landscape for jobs and income – particularly if industry is footloose and can quickly relocate to an area with lower regulation and resource costs. This does not bode well, environmentally nor socially, if responsible sustainable development is to be desired. Deregulation of planning laws (or similar directives) that protect out-of-town development may further produce environmental degradation and urban social problems if not controlled responsibly.

d. Services in education, health and crime prevention

In addition to quality housing and access to employment, further issues in urban areas will need to be analysed via an economic lens to ensure satisfaction of scarce resources is maximised without excessive detriment to the environment. Further social issues in urban areas that add further complexity to sharing space are those such as the desired provision of good education and health, and crime prevention.

Education by residents in urban areas can raise aspirations, dissolve socio-economic boundaries and enable an increase in income and wealth. If there is access to a good standard of education in urban areas the prospects for human development are enabled; therefore access to good public education will be critical to ensure all backgrounds have the opportunity to achieve their ambitions. The education system can face many inequalities within urban schools. One major issue is the extreme poverty in urban areas combined with a higher cost of living. Other factors include: overcrowded classrooms, outdated and meagre resources, run-down buildings and insufficient funding. As a result of these factors, quality teachers are discouraged from working in urban schools. This social problem is one to be overcome as it has been demonstrated that as well as maintaining dignity, it has been estimated (Denison, 1985) that education can create economic growth, with

20 per cent of the growth in output per worker in the US between 1929 and 1982 being associated with an increase in the general education of workers.

Improving health is another significant goal for urban areas to encourage sustainable development (for economic, social and environmental reasons). In economic terms, a healthier workforce will be happier, more satisfied and more productive. The opportunity for improved health in urban areas has some challenges. For instance, pollution levels from industry and traffic may increase the external cost in health value for those that reside in urban areas. Furthermore, inequalities in urban areas' health can be mirrored by wealth. For instance, on average the urban rich live for longer and in better health than the urban poor. For a squatter settlement in Manila, Philippines, nearly three times as many children die before their first birthday than in non-squatter areas and diarrhoea is twice as common and tuberculosis nine times as common. In the ('poor') Bronx, New York City, children are five times more likely to contract tuberculosis than their wealthier neighbours. In London, heart disease and respiratory diseases are twice as common in poor areas than in rich areas (WRI, 1996). The redistribution of wealth may therefore help re-balance health inequalities, as well as improvements in health within urban areas increasing economic opportunity. Access to quality healthy foods is also a consideration in the social problems of urban areas in the distribution of environmental resources. A particular problem of barriers to obtaining healthy food is the presence of food 'deserts', where the barriers may include lack of access to food retailers, availability of nutritious foods or a problem in the affordability of healthy foods.

A further social issue in urban areas that will affect the amount of resources available for allocation is the degree of crime-prevention provision. In terms of neighbourhood desirability and property value, where there is a high incident of theft and violent crime there will be a reduced neighbourhood demand and value. Crime is a particular social problem in urban areas, and also in the suburbs of those urban areas. Causality for crime can be considered in many instances for economic gain; for instance, theft and mugging may, in the first instance, be for some monetary gain – although the social reasons may be more complex, such as in instances of drug addiction. However, explanation for violent crime may go beyond conventional economic analysis. Poverty and lack of employment opportunities may also add some weight to reasons for higher crime in cities. Spatially, the higher concentrations of such crime can be attributed to those areas with large concentration of poverty, particularly those cities that have experienced rapid industrial restructuring and global shifts in sectors such as manufacturing. There is also further consideration that at a more micro level, those individuals living in poor areas may have a greater chance that they, too, will be poor – what has been termed 'neighbourhood effects'.

e. Transport and infrastructure (connectivity of urban spaces)

It is quite widely thought that the role of infrastructure, particularly transport, is one particularly useful way to understand the history of urban areas, particularly by looking at the methods used to transport people and goods inside cities (McMillen and McMillen, 2008). The importance of transport is nothing new, with Evans (1985) stressing that transport improvements appear to have been the primary cause of long-run changes in the structure of cities. Debate on transport in urban areas tends to focus on sustainable transport, with discussion centring on reducing the reliance and negative environmental and social impacts of private car use, promotion of public transport, and a consideration for a more 'walk-able' and 'cycle-able' urban fabric. As for the economics of transport, greater connectivity in urban spaces can provide greater physical mobility to help achieve better socio-economic opportunities, such as access to jobs and employment.

Access to a car for private transport can often improve mobility opportunities and thus improve job prospects. This means that low-car-ownership groups tend to coincide with lower income and vulnerable groups (e.g. elderly and disabled). Improved public transport and walkable opportunities to access employment are important for urban spaces. In improving transport more generally, social and economic benefits can emerge, such as reduction of road accidents, lower air pollution, increasing physical inactivity, reducing time taken away from the family while commuting, and reducing any vulnerability to fuel price increases – these are particularly improved if public transport options are economically viable and successfully implemented. Environmentally, less car use in cities would obviously reduce congestion and improve health factors related to improved air quality and less smog, as experienced by many international cities. It should be further noted that economic disadvantages resulting from congestion through transport blockages include wasting time and slowing the delivery of goods and services.

Urban areas in Europe can be highlighted as being areas that are seeking to tackle the issue of sustainable transport for mobility. It is recognised, for instance, in the EU Action Plan on Urban Mobility that cities need efficient transport systems to support their economy and the welfare of their inhabitants (EC, 2009). Furthermore, around 85 per cent of the EU 's GDP is generated in cities, and urban areas face today the challenge of making transport sustainable in environmental (CO_2, air pollution, noise) and competitive (congestion) terms while at the same time addressing social concerns. Urban mobility is also a central component of long-distance transport. Most transport, both passengers and freight, starts and ends in urban areas and passes through several urban areas on its way. It is therefore argued that urban areas should provide efficient inter-connection points for the trans-European transport network and offer efficient 'last mile' transport for both freight and passengers (EC, 2009). This demonstrates that transport mobility for increased economic and social well-being is an issue that needs to take into consideration connectivity both within urban areas (i.e. intra-connectivity of self-contained spaces) and the wider connectivity between urban spaces (i.e. city inter-connection).

Connectivity between spaces, as well as within spaces, is particularly important for understanding the allocation of resources. For some environmental resources the distribution will be via infrastructure channels that need careful strategic planning to allocate efficiently. Infrastructure for transport will entail the layout and construction or maintenance of roads, highways, rail, ferry routes and air-traffic paths that will have nodes or hubs that meet in (or near to) major urban areas (e.g. bus terminals, rail stations, airports, ferry terminals etc.). Further (inter- and intra-) urban infrastructure considerations are those such as utilities (such as water, gas, electricity, sewage), communications (such as telephone and internet fibre-optic lines) and green infrastructure (such as parks and wildlife corridors).

As similar to transport, efficient flows of utilities will enable maximum economic gains at the wider spatial scale. This will be due to economies of scale being generated for goods and services in infrastructure that would be considered a 'natural monopoly'; a natural monopoly being those goods and services that would be economically wasteful if there were many competitors operating in the market – for instance, there would be no wider economic advantage of having several rail lines going between the same place, or several bridges over the same crossing. The strategic use of green space in urban areas will also be able to add value to surrounding land use such as residential areas, or provide natural resource value by maintaining greenbelts or wildlife parks – not only to retain natural biodiversity but also to hold economic value of natural assets. Infrastructure as transport, utilities, communications and green space can therefore be costly externalities if development is permitted without regard for wider strategic planning. For instance, the building of a housing development in the middle of nowhere will have some cost external to the producer and consumer, either privately (e.g. as a management charge) or from public funds (to cover the external costs) in relation to building such infrastructure. Furthermore, economic value can be drawn from use of the planning system to ensure both private and public costs are minimised and benefits are maximised.

f. Poverty, social exclusion, disadvantage and inequality

The final social issues to consider in urban areas that can have an impact on resource allocation, particularly with regards to (in)equity of distribution, are issues of poverty, social exclusion and disadvantage. For several developed nations, and notably in the US, a large disproportionate incidence of poverty is concentrated among black-, aged- and female-headed households. Winger (1977) adds to this observation by stating the rural to urban differentials in poverty. He states that urban poverty is concentrated in the inner core of many central cities and that 'the urban ghetto is a fact and many of these places are where poor blacks live'.

Poverty and social exclusion

Contemporary study of 'poverty' and the emergence of the concept of social exclusion should firstly be introduced. Prior to the election of the first New Labour government in the UK in 1997, disadvantaged individuals and communities were those whose standards of living were lower than average and were generally also seen in terms of poverty. However, crude attributions that identified 'the poor' had begun to give way to subtler classifications. These initially grappled with a distinction between absolute and relative poverty. Although there are undoubtedly different interpretations, absolute poverty relates to a predetermined and universal (perhaps global) standard below which even the basic needs for food and shelter are not met. Whereas the latter is concerned with an agreed (low) standard of living relevant to a specific country or society at a particular point in time. For example, in the UK the working definition tends to indicate that poverty exists for those on lower than 60 per cent of the median income. Whilst the most important problem in the poverty discourses is with levels of income, some also make clear the links between poverty and its impacts upon health, life expectancy and access to services.

This differentiation in the definition of poverty connects with the newer concept of social exclusion. Originally used in 1970s France, relating to those with low incomes and few rights and opportunities in the employment market, since 1997 in the UK it has become a well-used term that can at times be difficult to pin down. Social exclusion is generally regarded as including poverty and disadvantage (and not the other way around), as well as aspects of life such as lower levels of, or the wrong sorts of, social connections or social capital, poorer job prospects and a reduced level of access to a wide range of services. However, the crucial emphases may vary. Indeed, one such emphasis has been upon a perceived spatial concentration of social exclusion. This is where area becomes important and, for the purposes of this topic, how poverty and social exclusion are simultaneously both cause and effect in urban areas.

Disadvantage in urban areas and 'area effects'

Images of 'rough' housing estates driven by social conflicts between different ethnic communities, with poor standards of housing and social isolation from the rest of society, continue to persist in many urban areas. Attracting some of these perceptions in the built environment is a long-running debate that concerns itself with the concentration of poverty and social exclusion within physically defined areas. Building on knowledge of the workings of markets and the economy from previous chapters, it is perhaps understandable that people with little income might live in lower-cost and lower-quality housing that is situated close together. Although this idea (called 'sorting') makes a certain kind of sense, there are suggestions that something else is going on in disadvantaged areas that impacts upon residents' life chances. These are the so-called 'area effects'. For instance, low levels of some kinds of social connections are not only compounded but exacerbated for those who live in such neighbourhoods – people become more isolated, confined to their estates and find it more difficult to access the 'corridors of power'.

However, finding evidence to support the area effects hypothesis has proved to be difficult. In part, this is due to complexity – that is, the acknowledgement that the interactions of individuals, communities, areas, services and wider socio-economic structures (at the regional, national and global scales) all work together to blur the picture. In addition, the relationships between parts of a neighbourhood, the neighbourhood as a whole and the neighbourhoods surrounding it are influential. It can certainly be argued that people are disadvantaged (and advantaged), socially and economically, by where they live. This, in part, is by birth; the other part can be improved by some form of changing the rules of the game. How this is done is the focus of area-based initiative (ABI) policy in the next chapter.

Summary

1   The Burgess concentric zone model is a 'city' model depicted as a series of five concentric rings rippling out from a central business district (CBD).

2   Economic explanations can be drawn from this concentric zone model in relation to rent price for land and transportation costs. With respect to land-rent costs, it would appear in the first instance that businesses and people with high incomes would tend to be closer to the centre.

3   The radial sector model by Hoyt described urban land patterns to be characterised as sectors (or wedges) running out like ribbons from the CBD at the centre – based on transportation links following railroads, highways and other transport arteries.

4   The multiple nuclei model as developed by Harris and Ullman (1945) demonstrates that similar industries with common land use and financial requirements are established near each other and are located around several centres or 'nuclei'.

5   Agglomeration economies, originally conceptualised as 'local industries', are where cost reductions can occur because economic activities are located in one place, and thus form economies of scale due to spatial concentration of particular industry. Agglomeration economies have four key classifications: (1) industry; (2) geographic; (3) temporal; (4) organisation of industry.

6   Urban sprawl radiates away from a central business district core. This sprawl can be somewhat extreme if drivers such as population (not always a key reason) and income continue to rise (enabling desires to drift further into the suburbs) or if regulation and directives in policy, such as via the planning system, curb and shape such horizontal urban expansion.

7   The analysis of housing should consider its multifaceted nature as a commodity that is a social good (to ensure needs are satisfied), a consumption good (for people to gain satisfaction from buying a house and living in it) and as an investment good (to buy in return for some form of financial return – such as rental income or asset growth).

8 Significant features particular to the housing market, and which hinder the efficiency (and thus failure) of the market are: (1) the time-lag effect that makes it difficult to physically supply housing in relation to demand and price changes in the short term; (2) the second-hand housing market is sensitive and responsive to the new building of stock; and (3) lack of competition in the supply of housing.

9 Demand for housing is related to the functions of population and income meaning that housing's relationship with income distribution will impact on its demand. One peculiarity of housing demand is the high cost of capital to purchase the 'product'.

10 Wider housing market considerations are despite the heterogeneity of housing supply and many characteristics of demand. The substitutability between the various categories of housing means that price changes in one market may have considerable impact on prices in other housing markets.

11 External market considerations are those factors that are external to the market (but could in some way be valued – e.g. by CBA) will be the reason for markets not being efficient in their allocation, and thus leaving housing either under- or over-allocated in the short to medium term.

12 Housing as a social good means that it can be viewed as a merit and public good beyond the market. Affordable housing at a decent standard is, arguably, not attained if left completely to the free market and therefore intervention in the market will have to occur if affordable housing at a decent standard is achieved.

13 Housing price and value is often linked to the neighbourhood characteristics in which the house is situated. A whole range of variables could therefore account for determining the price. What can be used in economics to test and research the value and price of property is via hedonic modelling.

14 For labour and employment in urban areas the connectivity of residents to their place of work will be significant to how urban spaces are developed.

15 The impact of changing global process have been varied in different cities depending on their previous development and the specific mix of employment sector and industry types – manufacturing, commercial, retail, office, finance etc. The fortunes of 'developed world' cities depend largely on their ability to respond to the challenges of globalisation.

16 Education by residents in urban areas can raise aspirations, dissolve socio-economic boundaries and enable an increase in income and wealth. In economic terms, a healthier workforce will be happier, more satisfied and more productive. Crime is a particular social problem in urban areas and also in the suburbs of those urban areas.

17 Debate on transport in urban areas tends to focus on sustainable transport, with discussion centring on reducing the reliance on, and negative environmental and social impacts of, private car use, a promotion of public transport and a consideration for a more walk-able and cycle-able urban fabric.

18 Infrastructure for transport will entail the layout and construction or mainte-
    nance of roads, highways, rail, ferry routes and air-traffic paths that will have
    nodes or hubs that meet in (or near to) major urban areas (e.g. bus terminals,
    rail stations, airports, ferry terminals etc.). Further (inter- and intra-) urban
    infrastructure considerations are those such as utilities (water, gas, electricity,
    sewage), communications (telephone and internet fibre-optic lines) and green
    infrastructure.
19 As similar to transport, efficient flows of utilities will enable maximum
    economic gains at the wider spatial scale. This will be due to economies of
    scale being generated for goods and services in infrastructure that would be
    considered a 'natural monopoly'.
20 Poverty can be conceptualised as absolute, relative and connected to ideas of
    social exclusion.
21 Disadvantage partly concerns itself with concentration of poverty and social
    exclusion within physically defined areas. Sorting, area effects and neighbour-
    hood effects (at a lower spatial scale, as similar to externalities and spill-over
    effects) can contribute to compounding disadvantage.

Chapter 12

Environmental resources and use in urban space

Bristol, UK

This chapter looks at 'environmental resource' issues in connection with the built environment. Externalities as the third-party costs and benefits are analysed first, using pollution as an applied example. This external cost is then examined in relation to the introduction of some sort of pollution control. Natural resource usage is then applied with regards to energy production and consumption in urban areas that have increasing wants for heat, power and light. Sustainable development objectives are further introduced in this chapter as both a theme and applying some applied examples in practice. For instance, with reference to sustainable energy resourcing, the relatively cheaper extraction cost of coal or fossil fuels provide short-term appropriation of economic wealth but with longer-term environmental degradation.

Valuing the environment as a resource will also be unearthed and applied in this chapter. An international theme will be explored that considers resource allocation at the global level. International co-operation and agreement on environmental issues require economic thinking on both local and global scales, especially as environmental problems have less regard for the national administrative boundaries they cross. Climate change and global warming are the most prominent of international and global concerns and are receiving attention from both policy and economic thinking. Human-caused release of CO_2 is contributing to global warming and climate change and should be ignored at our peril. The economic consequences of climate change such as extensive flooding and sea-level rise will have economic consequences, and not necessarily are caused by the country being affected by climate change.

a. Externalities

As discussed in Chapter 7, externalities can be more broadly defined as the spillover or third-party effects arising from the production and/or consumption of goods and services in the marketplace. In order to improve capture of full costs in using environmental resources that are produced or consumed within urban areas, models can be used to conceptualise and value the external costs in using environmental resources. The degree of cost will depend on the environmental sustainability in the production and consumption of the good or service. A more sustainable good or service would be one that has less external costs at a higher volume of output.

Socially optimal level of output for energy

The socially optimal level of output model (Figure 12.1) considers the assimilative capacity of external costs, such as pollution costs, to determine the point at which resource costs stop being assimilated by the environment and society by beginning to detrimentally impact upon it.

In using Figure 12.1 as a model, and the example of a coal-fuelled power station, at a global level the optimum amount of coal used in the power stations that can be assimilated and absorbed by the planet are at quantity Q1, with a level of pollution (seen here as CO_2 volume) emitted at a level of A1. This point is regarded as the socially optimal level of output – any further output would be directly affecting society through environmental degradation. If any further coal is used to fuel energy production beyond Q1, this would take the amount of pollution (as CO_2) beyond the finite assimilative capacity (expressed as a straight line) of the planet, and as a result generate a surplus external cost that is the difference between the total released pollution and the assimilative capacity.

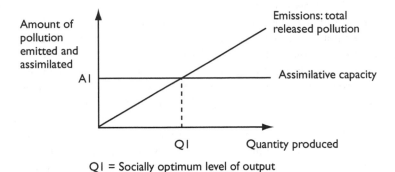

QI = Socially optimum level of output

Figure 12.1 Socially optimal level of output
Source: Author

b. Pollution control (of environmental resources at a market or zero price)

The firm could internalise some of these external costs of pollution if some intervention is applied by governing authorities. This means that incentives provided within the marketplace can alter and direct change in production and consumption patterns. For environmental goods and services, price incentives centre around restricting and regulating production and consumption, as at their most basic, the price of inputs for the resources (without extraction and processing costs) are zero. Environmental resources are, in essence, zero-priced or free goods and services, such as the price of fresh drinking water (prior to collection and distribution) being free at source (depending on the ownership of the land from which it is being sourced). Without market incentives and regulation, an *unfettered* or uncontrolled price mechanism will use too much of zero-priced goods. In the fresh water example, a zero source cost of water will mean an over-production of the product if it is to be turned into a commodity and sold, as the cost to produce is a non-controlling factor, or more simply, the zero price provides no incentive to produce less of the good. If a surplus and over-supply of free goods is produced there will be a cost to society above the private internal cost to the firm producing the fresh water for consumption. This excess cost and divergence away from the private cost is referred to as the 'social cost'.

Market inefficiency of zero-priced goods can be reduced if individuals and organisations control their behaviour though some non-price incentives. Behaviour can be considered a non-price determinant of demand function with regards to changes in tastes and preferences altering the level of demand at the same price point. An example of non-priced determinants of demand that alter the behaviour (tastes and preferences or fashion) of zero-priced goods is through education. Unfettered markets could have improved environmental quality if consumers change demand through education, awareness and choice of less-polluting products. Other non-price determinants have been covered in Chapter 6, and will enable thinking on the incentives and guidance that can be directed to zero-priced or free environmental goods and services.

Incentives to environmental goods and services can therefore shape markets or correct market failure. These incentives can be viewed as either direct or indirect in the way in which they operate. Complete allocation control to the free market can be created via direct full privatisation. In the UK and in most other countries, the supply of (free) water utility has been in complete control of the government. In a process of privatisation this relaxation of governmental control has meant that competing water companies could form by leasing parts of the infrastructure and selling water as a commodity at a price determined by the quantity of units consumed by the end-user. The movement in viewing water provision from a public good to one that is private mirrors the transformation of value of the environmental resource as zero-priced to one that has a price attached to it. Other direct regulatory changes could see a shift in the opposite direction from a deregulated free market to that of a command-and-control approach to environmental distribution. For instance, the government approach could be to decide what can and cannot be produced and polluted with respect to the environment. For instance, environmental standards can be directly commanded and enforced via legislation; such as setting and enforcing water purity standards for firms that are operating near water outlets. As well as direct regulation as an enforced incentive to value environmental resources, indirect incentives are another mechanism that can alter their functioning. Indirect incentives could be regulations such as taxes and charges that financially incentivise the value and allocation of environmental goods and services. A higher service charge for water for residential use may discourage the quantity of water used by the households subject to the service charge. As a result of such indirect manipulation of the market, this will adjust the price in existing markets for environmental goods and services.

A radical intervention into the market for environmental goods and services and which will change their economic functioning is by creating 'new markets'. Governments can create new markets for environmental services by direct regulation and creation of new tradable entities that act as a new 'currency' in the marketplace. A classic example is the creation of environmental carbon credits that are traded and regulated. Carbon emissions trading schemes are a market-based mechanism for helping mitigate the increase of CO_2 in the atmosphere. Carbon trading markets are developed when they bring buyers and sellers of carbon credits together with standardised rules of trade. For instance, entities that manage forest or agricultural land might sell carbon credits based on the accumulation of carbon in their forest trees or agricultural soils. Similarly, business entities that reduce their carbon emission may be able to sell their reductions to other emitters. Critique of this carbon credit market is apparent; this system may inhibit developing nation's economic development as they have fewer financial resources to afford the permits necessary for developing an industrial infrastructure. That is if the standard next development phase must follow modes of production in a linear process of agricultural, industrial, service and creative modes.

Pollution control: charges, permits and deposits for environmental use

Indirect economic mechanisms can therefore shape and influence the market price and resultant value of environmental resources that are used and produced in urban areas. Table 12.1 consolidates five key instruments that can be used to influence the price of environmental resources on the market. Firstly, emission charges can be used to ensure that the emitter has to pay for any polluting activity, so with the charge they are encouraged to pollute less and thus be charged less. The charge is often related to the quantity (and quality) of pollutant damage. So in the case of water pollution by a firm in proximity to a water source, the cubic feet of water polluted could be charged, plus the intensity of pollution effecting its regenerative capacity to return to pure water will equally be charged more intensely.

Secondly, user charges are another instrument to influence price of environmental goods. As different to emission charges, a user charge is one where the more of a resource is used the greater the charge allocated. For instance, the more units of coal that are used in producing energy in a power station the greater the charge is attached to the use of coal.

Product charges are the third instrument, and are based upon the particular product that is being produced to gauge how much pollution is caused, and hence charged by a regulatory force. The harm in producing cars may be more polluting than the creation of bicycles (if compared at equivalent value, e.g. one car and ten bicycles), and therefore the charge will be more heavily felt in the former product.

Table 12.1 Five Instruments using charges to price environmental use

Instrument	Descriptor
1. Emission charges	Charge on discharge of pollutants into the air, water, soil or as noise – related to the quantity and quality of pollutant damage
2. User charges	For revenue-raising in relation to treatment, collection, disposal and administrative costs
3. Product charges	Harmful environmental products have charges attached when used in production, consumption or disposal
4. Marketable permits	Permits given that match an environmental quota or allowance for pollution levels; these can then be traded
5. Deposit-refund systems	Deposit paid on a potentially polluting product; authorities return deposit (or part of the deposit) depending on the level of pollution on the returns

Source: Author

Marketable permits are the fourth instrument and draw on the carbon emissions trading example discussed earlier. Here a certain quota of carbon emissions for a firm or nation can be set and any production under the set quota will generate a surplus carbon credit that can be traded to firms or nations that over-produce goods and services that emit carbon.

The fifth and final instrument that can affect the price of environmental resources, particularly in affecting price in controlling their pollution discharge, is the use of deposit-refund systems. A deposit-refund system is where a firm pays a deposit up front to an authority if they are potentially likely to emit pollutants in their process. At a certain point in the production process, a clean cut-off would be at the end of the process. The authority will return the deposit if no pollution has occurred, or retain part of the deposit if there are partial pollution emissions. An example could be a logging firm that pays a deposit to cut down a certain number of trees, and if at a point of audit they have over-logged, they will lose part of the initial deposit.

Pollution control example: carbon tax

A specific and simple indirect regulatory approach to influencing market behaviour on environmental pollutants through price is via a tax. Here the mechanisms that occur through a carbon tax are now demonstrated. The unsustainable release of carbon is regarded as a type of pollutant, particularly as the excessive release of carbon in the extraction and burning of fossil fuels will produce an atmosphere that retains heat and thus contributes to global warming and climate change. To reduce global warming through lower release of greenhouse gases (particularly CO_2), a tax on extraction or consumption of carbon fuels may be a strong incentive that is fed through into the price mechanism. More complex mechanisms can be attached to taxing carbon; for instance, a carbon tax could be increased at a graduated rate if the carbon content of fuels increases. As coal has higher carbon content than gasoline, it could be taxed at a higher rate. Furthermore, gasoline has a higher carbon content than natural gas, so gasoline could be taxed at a higher rate than natural gas.

The influence on the market from a carbon tax is not straightforward. For instance, a graduated tax depending on the level of CO_2 emissions does not necessarily mean that the electricity sector would alter its source of fuel for energy to those that are less polluting. The effect is also not entirely clear because the impact of a tax will affect the consumer as well as the producer, and it is the consumer that may substitute to a less-expensive (less-taxed) and less-polluting heating system in their households; meaning a substitution from say a coal fire to gas central heating, assuming households can easily switch. Further complication in controlling price arises in that all energy-using sectors are increasing prices whilst consumers are being more energy conservative, either through affordability problems or by becoming more environmentally conscious and aware. The result is an environmentally sensitive shift to less CO_2 polluting energy sources such as natural gas and renewable sources.

As well as these complexities associated with a carbon tax (as an example of indirect taxes as an instrument of environmental resource use), there are very clear disadvantages associated with a tax instrument being used to affect the market for potentially polluting environmental resources. One disadvantage of a carbon tax is that it is socioeconomically regressive in terms of economic development. This is because the poor will pay a larger percentage of their income in carbon tax, and inversely, wealthier social strata will pay less of a proportion of their income on carbon tax. Secondly, a carbon tax may be detrimental as the tax adds to the tax base of the taxing country, and hence paradoxically, may encourage the use of carbon fuels as public revenue can be extracted from it. Thirdly, difficulties of carbon taxes are that the outcome achieved in the taxing country may not be achieved in another country, which may be emitting an extremely high level. This will mean that, on a global scale, if there is not unilateral agreement between countries there will be a negligible impact on global warming and climate change.

c. Natural resource economics

Resources are the backbone of every economy. In using resources and transforming them, capital stocks are built up that add to the wealth of present and future generations. However, the dimensions of our current resource use are such that the chances of future generations (and the developing countries) to have access to their fair share of scarce resources are endangered. Moreover, the consequences of our resource use in terms of impacts on the environment may induce serious damages that go beyond the carrying capacity of the environment (EC, 2011). Natural resource economics (NRE) is a branch of economics that is of interest in this field, and is one that can be tailored to suit interests involving the use and management of shared urban spaces. Many of the concepts and techniques relating to natural resource economics have been covered in the preceding chapters of this book. However, it is worth outlining all elements to ensure that they can be explored in more detail.

As an introduction, natural resource economics deals with the supply, demand and allocation of the Earth's natural resources. One main objective of natural resource economics is to better understand the role of natural resources in the economy in order to develop more sustainable methods of managing those resources to ensure their availability to future generations. Natural resource economists study interactions between economic and natural systems, with the goal of developing a sustainable and efficient economy. Natural resource economics aims to address the connections and inter-dependence between human economies and natural ecosystems. Hence this field tends to draw closely with those ideas mapped out in ecological economics (EE) as discussed in Chapter 2. For instance, economic connection and modelling is made with fisheries, forestry and minerals (i.e. fish, trees and ore), air, water and the global climate. Value in NRE has also moved into the management and value of recreational use as well as commercial use of resources. Of particular interest to NRE is the understanding via an economic lens as to differences between perpetual resources and exhaustible resources – what have also been referred to as renewable and exhaustible resources in Chapter 2 when discussing difference between ecological economic (EE) and environmental resource economic (ERE) paradigms.

For urban areas and natural resource economics, several features can be applied. For instance, natural resources available within and used by urban areas will have an impact on its economy, and thus develop more sustainable methods of managing those resources. Cities with close proximity to natural resources may have advantages such as low transportation costs of water availability; although the strength of the urban economy may enable it to command a greater consumption of natural resources from locations that have no direct ecological and environmental connection to the urban space under investigation – such as wealthy cities enabling transportation of water over great distances. Green city and ecocity provision (Chapter 13) are exemplar of a shift to renewable energy (and utility, transport, infrastructure, housing, industry, commerce, services) provision within urban spaces; although the use as well as management and allocation of natural resources may be less sustainable economically and socially if the urban space operates in isolation to other functioning urban spaces.

d. Energy economics

Energy use in urban areas is a particularly important theme as greater urbanisation means that the energy needs will increasingly be consumed in shared urban space – in private and public property and within the spaces in-between as the public realm (e.g. energy for street lighting). Focus on energy use and management in urban spaces is introduced in another applied economic sub-field referred to as energy economics. Energy economics concerns itself with issues related to forecasting, financing, pricing, investment, taxation, development, policy, conservation, regulation, risk management, insurance, portfolio theory, fiscal regimes, accounting and the environment.

By definition, energy economics studies forces that lead economic agents – firms, individuals, governments – to supply energy resources, to convert those resources into other useful energy forms, to transport them to the users, to use them and to dispose of the residuals. It studies roles of alternative market and regulatory structures on these activities, economic distributional impacts and environmental consequences. It studies economically efficient provision and use of energy commodities and resources, and factors that lead away from economic efficiency. Energy commodities of interest would be those such as gasoline, diesel fuel, natural gas, propane, coal or electricity, which can be used to provide energy services for human activities such as lighting, space heating, water heating, cooking, motive power and electronic activity. Energy resources – e.g. crude oil, natural gas, coal, biomass, hydro, uranium, wind, sunlight or geothermal deposits – can be harvested to produce energy commodities (Sweeney, 2000).

e. Resource valuation and measurement

Approaches for valuation and measurement in urban and environmental economics are both positive and normative. These approaches will apply certain techniques in order to measure and value resources, and it is these techniques that will be introduced and explained. Key measurement and valuation techniques are: asset-based techniques; contingent or stated preference (rather than revealed preference based on market price) models using willingness to pay (WTP) and willingness to accept (WTA) principles; replacement costing; market pricing; and environmental charging instruments.

Valuation in positive and normative economics

Economic costs and benefits will have different approaches depending on whether a positive or normative economic approach is taken. Positive economics is an approach that explains economic activity in a more factual and testable manner. Positive economics is more aligned to the natural sciences by appropriating disciplines such as mathematics in order to precisely attribute relationships and correlations between various components and factors. Using such a 'hard science' approach, positive economics attempts to describe *what is, what was, or what will be* in respect to time. This enables current, past and future economic activity to be retrospectively analysed for improved foresight in decision-making. Within positive economics the emphasis on facts means that disagreements in arguments, particularly over scarce resources and infinite wants can be resolved by an appeal to the facts. For instance, if it is argued that an urbanising geographical space consumes more energy and emits greater levels of CO_2, the addition of statistics to evidence that urban areas consume and release more CO_2 compared to rural areas will be of prominence. Positive economics is also an economic approach where statements of 'truth' can be tested. As per the urbanising energy-use example, facts revealing that there is greater energy use in urban areas can be tested by recording and measuring, and then recording, how much energy is consumed by all of the households within both urban and rural neighbourhoods.

Distinct to positive economics is 'normative economics'. Rather than provide testable facts as per positive economics, normative economics is more interested in dealing with what *ought to be* in society. It is thought in this approach that even if tested evidence provides some universal 'truth', it does not necessarily mean that the 'fact' is of merit to progress in society. For instance, the high correlation of higher energy use per unit of space in urban areas does not necessarily mean that all urban areas should be deurbanised if CO_2 emissions are to be arrested. The 'should be' is more of interest to normative economics where any disagreements, such as arresting the amount of CO_2 release, can be resolved by negotiation. Any negotiation will involve an element of value judgements being traded between parties in order to reach a clearer 'truth' through dialogue. This normative economic approach can therefore provide guidance on desirability of a programme before or after implementation; such as whether CO_2 releases in urban areas are the key contributor to climate change and global warming, and therefore lead to justification that an urban CO_2 reduction programme should be in operation.

Positive and normative economics are not always diametrically opposed approaches, as aspects of both approaches can help to determine an understanding of what economic forces are operating on a particular event. Both applications are useful for urban and environmental economic analysis. If the relationship between inter-urban trade (e.g. individual and organisational exchange between cities) and the environment is considered, both positive and normative economic approaches will provide some input. Positive economics in city trade would be more interested in describing impacts of trade on the economy and environment. For example, if a higher gross value added (GVA) was attained for a city, what environmental impact would occur and at what correlation value? The difference for normative economics in this example is that it would generate judgement and produce guidance as to whether trade was desirable in the first instance no matter what environmental impact could be calculated.

The environment as an asset

So there are different approaches to economics, as positive and normative, and in considering these approaches differing measuring and valuation techniques are used in urban and environmental scenarios. The first technique to look at is an environmental asset approach that can be used to explain and quantify phenomena by urban areas. An asset approach to resources is one where something tangible has value now as it has the potential to generate future economic benefits. The asset approach also means that its current asset value has been provided by some past economic activity. It is also an asset resource that is controlled by an entity with respect to the result of these past transactions or events from which future economic benefits have been obtained. If 'life' itself is seen as a future benefit that can be generated from the environment, the environment is, as a result, an asset that provides future life-support services. As such, in thinking of the environment as a measured and valued asset, any future fall in the future life-sustaining services will be reducing the asset value of the environmental entity. Inversely, it is therefore important, if the goal is to protect the environment (as per one of the main macroeconomic objectives), that individuals and groups try to prevent depreciation of the asset so that it can continue to provide aesthetic and life-sustaining services.

So to value the environment as an asset, measurement of its future benefits need to be conceptualised and recorded. In doing so, thought and questioning need to be focused on what the environment provides as benefit (and cost) for the economy into the future. Firstly, future benefits by the environment are natural resources as raw materials that are transformed into goods and services in the production process. The future benefits of the environment as an asset are therefore the economic benefits of the goods and services produced from the natural resources, such as the future economic benefits of deforestation to provide wood as a building material that will last X amount of years; although this will come with a future opportunity cost, that of providing economic benefit if it was not extracted for processing and therefore of future value; or as providing future value in preventing subsidence and retention of soil nutrients; or the provision of shelter and habitat in the case of forests.

As well as environmental assets being valued due to their future economic benefits as raw materials, future energy units derived from the environment will enable the environment to be valued as an asset. Energy, for instance, fuels the production transformation process and therefore has future economic benefit potential – and therefore energy derived from the environment means that the environment has asset value. The environment also has asset value as it can provide direct services to consumers for future benefit. Economic goods and services from the environment to be consumed in urban areas (and rural areas) are those such as clean air and water for a more productive workforce, the nourishment from food to enable future productivity, and/or the environment can have asset value in that future benefits are extracted from shelter and clothing over several years.

Environmental asset values can appreciate (go up in value) or depreciate (go down in value) within a closed internal economic system. Otherwise, the value is externalised as goods and services waste, and needs to be valued as an externality. The environmental asset depreciates as the future benefit starts to generate fewer future benefits if there is a greater output of waste. An excessive provision of waste depreciates the environmental asset at a point where the absorptive capacity of nature is exceeded. To illustrate, a paper mill that does not use sustainable wood sources in its production process will risk the creation of infertile soils that do not allow any further cultivation of trees for future paper-use benefits or other wood-derived goods and services.

Contingent valuation (stated preference): willingness to pay (WTP) and willingness to accept (WTA)

Environmental assets introduce the idea that economic concepts such as assets and revenues can be attributed to environmental resources and urban spaces. In addition, it has been intonated that asset approaches can begin to measure and value such concepts. Another valuation method in urban and environmental economics is through the willingness to pay (WTP) technique. WTP is a standard measurement of benefit in economics, and is, as the term suggests, what an individual or group is willing to pay for a particular good or service that is not directly sold on the market. The WTP is the price, sacrifice or exchange that a person will enter into in order to avoid a detrimental consequence, such as incurring pollution or a reduced quality of life. An example of a WTP value would, for instance, be the a £5 charge for visitors to enter a national park, and as a proxy, provide some value as to the national park's economic worth, in addition to the donations and public funds directed towards its conservation and preservation.

As it is a willingness to pay, the WTP concept has strong connection with demand rather than supply. Supply connections in this type of valuation method would be a willingness to accept (WTA) technique. The connection to demand in WTP therefore means that it is the purchaser that has a WTP that is equal to or exceeds the price. This means that the consumer has a WTP at a particular maximum (rather than minimum) value. To attach some illustration, if a high-speed rail link was going to be constructed to free up traffic congestion in the centre of an urban area, an economic analysis using WTP could be attached. The WTP for such a high-speed rail scheme would be to generate what the average price would be that commuters would be willing to pay for such a scheme in order to improve quality of life and speed up the commute and leisure travel times. If, as an average, the urban areas' actively travelling population (of say 50,000 people) were willing to pay a minimum of £5 per day and a maximum of £10 per day to use the service, the WTP principle would suggest a charge of £10 and an estimated annual valuation of the scheme at £182.5 million (10 × 50,000 × 365). Note that WTP is constrained by an individual's level of wealth (a function of demand). If applying WTP to quantifying one's own life, the WTP to not suffer the detriment of death would be the maximum amount of wealth the individual owns (assuming that all beings wish to survive, notwithstanding suicide). Conversely, then, the WTA life would be an accepted value at any price (again, notwithstanding suicide) if all people wish to live irrespective of the 'price tag'.

Replacement cost approach

Another economic technique used to value environmental (built and natural) resources is the replacement cost approach (RCA). The RCA operates in a similar way to the opportunity cost principle, with opportunity cost being the value of the next best opportunity forgone. For instance, the value of a wetland is the opportunity cost of it being a bypass, a housing development or an airport – or whatever has the highest cost to build. Similarly, the RCA is a measure of the benefit formed through avoiding damage as a result of improved environmental conditions. For instance, the asset value of property close to the waterfront will benefit in value if flood resilience and resistance measures are taken to counter sea levels rising in response to global warming. As a specific measurement of replacement cost, an approximated market value can be attached in terms of the cost to prevent, restore or replace the damage in question. In the flooding and property example, the cost to restore flooded property will be the value, and therefore the aggregate cost of all potential flood replacement costs, of the environmental phenomena in question.

Replacement cost approach in the urban area and in relation to real estate is therefore a valuation approach to real property based on what it would cost at current prices to build an equivalent new structure to replace the old one (that was built in a previous time in an entirely different cost structure). In applying another environmental incidence causing replacement of a built structure, the occurrence of acid rain due to changes in vegetation dynamics from over-cultivating can be valued using RCA. Acid rain deteriorates a nation's infrastructure, such as highways, bridges and historical monuments, and therefore has a cost to infrastructure if damage occurs; and also the reduction in acid rain can, as a consequence, have a benefit value. If the replacement cost of repairing built structures from acid rain is $1 million dollars, the benefit value of reducing acid rain to a level where no damage is incurred is therefore also $1 million. If a bill is passed by government to reduce sulphur and nitrate emissions by 50 per cent, this would mean that from the replacement cost valuation of built structures, a benefit of 50 per cent of replacement cost, in this example $0.5 million, is attributed to sulphur and nitrite emissions reduction. In short, replacement cost is the current market-value benefit of reduced costs in restoration or replacement damage of entities that can be valued, such as real estate and infrastructure. The valued benefits of reduced damage from environmental emissions as pollution are the savings realised from reduced expenditure on repairing, restoring and replacing the nation's valued goods and services such as infrastructure and property.

This abstract costing concept has some difficulties despite its ability to measure and value entities that may be described as intangible. Firstly, the environmental damage by 'abnormal' flooding or acid rain onto structures may not be able to be completely repaired or replicated. For instance, a historical monument has artistic value that may not be able to be (a) measured, or (b) be restored to its original form. If this replacement cannot be completed it cannot be given a full cost, and thus the benefit of reducing the environmental damage cannot be quantified. Even if a replica is put in the place of the original structure at a certain cost, the replica may have less worth than the original. As a result of this incomplete replacement, the replacement cost approach is a tool to consider for economic analysis of urban areas, particularly in the built environment that constitutes urban areas, but should be used with caution and expertise as to whether the full replacement cost has been attributed.

The market pricing approach (revealed preference)

Cost, benefit, price and value can now be seen to be formed from an abstract of economic concepts rather than a simple acceptance of a number being applied to an environmental (natural or built) entity. As seen in Chapter 6, the standard approach to deconstructing what is meant by price is the use of the market of a good or service that incorporates a combination of supply, demand and quantity. The environment as a resource is no exception and the market mechanism can be applied to provide an equilibrium price for environmental resources. One way to measure the environment in relation to market values is the way in which environmental improvement, and thus the reduction in externalities, means that the cost and benefits of environmental improvement is internalised into the market and causes increase or decrease in real inputs/outputs. For instance, this increase in output from environmental improvement will mean that the value of a good or service increases, and thus the gap between what it was worth before and after environmental improvement is the added value. To solidify this concept, an increase in crop yield due to higher air-pollution controls means that the reduction in pollution as a result can be quantified – in terms of what the increase in crop yield is.

In returning to basic neoclassical (and thus classical) market diagrams, this valuation of environmental pollution can be shown. In Figure 12.2, the increase in crop yield from improved air quality means that the supply of crops can be increased, as demonstrated by a shift in the supply curve. The previous revenue would have been represented by the rectangle formed by P0Q0; but now with increased production from environmental improvement the new revenue is Q1P1. The difference in these two revenues, before and after, is therefore the revenue gained from the increased yield and thus the value that can be attached to pollution control.

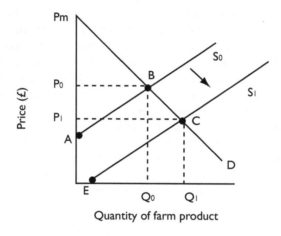

Figure 12.2 Supply increase and revenue changes in crop yields to measure pollution control

Source: Author

f. International environmental issues and development

International perspectives are paramount when studying the allocation of natural resources within and between shared urban spaces. Environmental policy in response to environmental problems are seen to have developed at an international platform for two key reasons: (1) many of the world's threatened natural resources are shared resources or 'common property' (such as the oceans); and (2) because pollution travels – as actions in one part of the world will affect the quality of life in another part of the world (Turner *et al.*, 1994). For environmental economics, difficulty may arise if a pollution control is implemented in one country when it is several other countries that are contributing to pollution. In terms of governance, there may not be a transnational authority with the power to impose a tax unilaterally.

It is widely thought that poverty is the greatest 'cause' of environmental degradation. For instance, the poorer someone is, the less likely they are to worry about tomorrow, with the immediate concern of food today. This suggests that poorer communities will show little concern for sustainability and will not undertake conservation practices to prevent soil erosion, or plant trees. It is true that poor people are often trapped in just such situations. But it is not so much that they do not care about tomorrow, but that they have limited ability to do anything about conservation if it diverts resources from the process of meeting today's needs. Nor is it universally true that poor communities fail to conserve resources. Many have fairly elaborate structures of rules and regulations that are aimed at conserving resources. As such, initial solutions may rest on rights and land resources that must be clearly defined to enable monitoring and regulation.

The urban poor, who are unable to compete for scarce resources or protect themselves from harmful environmental conditions, are most affected by the negative impacts of urbanisation. The growth of large cities, particularly in developing countries, has been accompanied by an increase in urban poverty, which tends to be concentrated in certain social groups and in particular locations (UNEP, 2002). As for the environment, in cities with human and capital resource imbalances (i.e. developing country urban areas), further impacts are experienced. The use of biomass fuel also causes indoor and outdoor air pollution. Other effects can be felt further afield such as pollution of waterways, lakes and coastal waters by untreated effluent. Air pollution from cities has an impact on residents' health as well as on vegetation and soils at a considerable distance. Urban transport contributes to air pollution and the large concentration of cars and industries in cities causes the lion's share of urban global greenhouse gas emissions.

As demonstrated by UNEP (2002), cities are often located in prime agricultural areas. If this land is converted for urban uses, this puts additional pressure on nearby areas that may be less suitable for agriculture. Urbanisation in coastal areas often leads to the destruction of sensitive ecosystems and can also alter the hydrology of coasts and their natural features such as mangrove swamps, reefs and beaches, which serve as barriers to erosion and form important habitats for species. Water is a key issue in urban areas. The intensity of demand in cities can quickly exceed local supply. The price of water is typically lower than the actual cost of obtaining, treating and distributing it, partly because of government subsidies. Another problem emerging in urban areas affecting development is the lack of suitable landfill sites to cater for the increasing demand for solid waste disposal. Poor sanitation creates environmental and health hazards, particularly by direct exposure to faeces and drinking water (ibid.).

g. CO_2 release, climate change and global warming

Shared urban spaces are platforms that consume and produce carbon dioxide and thus contribute to global warming and climate change. As noted in a paper by IIED (2007), for those cities affected by climate change there are also tens of millions of people in low- and middle-income nations whose homes and livelihoods are at risk from sea-level rise and storms, although they have made very little contribution to global warming. The economic cost of losing certain cities for which adaptation costs are too high may be relatively small for national economies. Adaptation will be easier and cheaper if greenhouse-gas emissions are reduced – so both the amount of adaptation and the rate at which it must be implemented are lessened. Adaptation plans must also bring benefits to the billion urban dwellers currently living in very poor-quality housing, in tenements, cheap boarding houses and illegal or informal settlements. These billion people include a large part of the population whose homes and livelihoods are most at risk from climate change. It is further argued in the paper (ibid.) that technology-driven, market-led responses to climate change does little for them. The key issue is how to build resilience to the many impacts of climate change in tens of thousands of urban centres in low- and middle-income nations.

In very basic economic terms, the economic cost of climate change will mean that it will remain on the international (and urban) agenda for the foreseeable future. With the Stern review (Stern, 2006) as a point of departure in tackling climate change the economic input is clear with regards to understanding the problem and mitigating its effect. Despite some critique, it is stated that climate change is the greatest and widest-ranging market failure ever seen, presenting a unique challenge for economics. Furthermore, it provides prescriptions including environmental taxes to minimise the economic and social disruptions. The main conclusion is that the benefits of strong, early action on climate change far outweigh the costs of not acting. The review points to the potential impacts of climate change on water resources, food production, health and the environment. In more detail, it is argued that, without action the overall costs of climate change will be equivalent to losing at least 5 per cent of global gross domestic product (GDP) each year, now and forever. A wider range of risks and impacts could increase this to 20 per cent of GDP or more. In terms of recommendations, it is proposed that 1 per cent of global GDP per annum is required to be invested in order to avoid the worst effects of climate change (Stern, 2006). This economically 'leaning' review sits as part of a wider environmental policy that aims to address many contemporary challenges that are demonstrated in Chapter 13.

Summary

1   The socially optimal level of output model considers the assimilative capacity of external costs, such as pollution costs, to determine the point at which resource costs stop being assimilated by the environment and society by beginning to detrimentally impact upon it.

2   Environmental resources are, in essence, zero-priced or free goods and services. Without market incentives and regulation, an *unfettered* or uncontrolled price mechanism will use too many zero-priced goods.

3   Market inefficiency of zero-priced goods can be reduced if individuals and organisations control their behaviour though some non-price incentives.

4   Incentives to environmental goods and services can shape markets or correct market failure. These incentives can be viewed as either direct or indirect in the way in which they operate.

5   A radical intervention into the market for environmental goods and services, thus changing their economic functioning, is by creating 'new markets' such as carbon emissions trading schemes.

6   Five key instruments that can be used to influence the price of environmental resources on the market: (1) emission charges; (2) user charges; (3) product charges; (4) marketable permits; (5) deposit-refund systems.

7   A tax on extraction or consumption of carbon fuels may be a strong incentive that is fed through into the price mechanism. More complex mechanisms can be attached to taxing carbon. For instance, a carbon tax could be increased at a graduated rate if the carbon content of fuels increases.

8   Complexity in outcomes involves: (1) a graduated tax depending on the level of CO_2 emissions does not necessarily mean that the electricity sector would alter its source of fuel for energy to those that are less polluting; (2) the consumer may substitute to a less-expensive (less-taxed) and less-polluting heating system in their households; (3) complication of controlling price is made in that all energy-using sectors are increasing prices; (4) it is socioeconomically regressive in terms of economic development; (5) a carbon tax may be detrimental as the tax adds to the tax base of the taxing country; (6) the outcome achieved in taxing country may not be achieved in another country.

9   Natural resource economists study interactions between economic and natural systems, with the goal of developing a sustainable and efficient economy. Of particular interest to NRE is the understanding via an economic lens as to differences between perpetual (renewable) resources and exhaustible resources.

10 Energy use in urban areas is a particularly important theme as greater urbanisation will mean that the energy needs demanded will increasingly be consumed in shared urban space. Energy economics studies forces that lead economic agents – firms, individuals, governments – to supply energy resources, to convert those resources into other useful energy forms, to transport them to the users, to use them and to dispose of the residuals.

11 Resource valuation and measurement approaches are both positive and normative. Key measurement and valuation techniques are: asset-based techniques; willingness to pay (WTP) and willingness to accept (WTA) principles; replacement cost approach (RCA); market pricing; and environmental charging instruments.

12 Positive economics is an approach that explains economic activity in a more factual and testable manner. Positive economics is more aligned to the natural

sciences by appropriating disciplines such as mathematics in order to precisely attribute relationships and correlations between various components and factors.

13 Environmental problems are seen to have developed on an international platform for two key reasons: (1) many of the world's threatened natural resources are shared resources or 'common property'; and (2) pollution travels.

14 The economic cost of climate change will mean that it will remain on the international (and local) agenda for the foreseeable future. The benefits of strong, early action on climate change far outweigh the costs of not acting.

Chapter 13

Policy and contemporary challenges

Los Angeles, USA

This final chapter reveals more applied applications of urban and environmental economics. There is a brief overview of urban policy more broadly followed by an outline of contemporary environmental policy. For urban and environmental economics, the combination of both policy strands are drawn together to demonstrate more holistic policy progression that incorporates sustainable development – and for the attention of urban areas, sustainable 'urban' development. Applied techniques are also explained and critiqued within the narrower use in practice of appraisal, assessment and evaluation. More technocratic introduction to the development of informatics demonstrates how such techniques have become embedded in sophisticated analysis for better decision-making. As a conclusion, but with consideration for the future, current events are given in order to demonstrate the forthcoming issues for shared urban spaces given the environmental resource challenges that are becoming more prevalent.

a. Economics in urban and built environment policy

Urban policy

To get some sort of understanding of urban policy, various scales of schemes, initiatives, projects and programmes need to be outlined. The sheer number of policies, even in the last 50 years, is high, and as they are often specific to a particular national government, the country-specific urban policies make an overview of all urban policies around the globe too lengthy for this book. However, a brief introduction to the varying types of policy (e.g. area-based initiatives and property-led initiatives) are given, with particular weight attached to the UK and US respectively.

If we outline some policies that try to address poverty and social exclusion manifested at smaller scales (such as those at the neighbourhood level), authorities have taken policies that are directed towards areas as well as people directly. For the UK, these ABIs have taken various approaches and appear to have increased during the New Labour period (1997–2010) – examples include Sure Start and the equally long-running New Deal for Communities (NDC) programme. Linked to some critique of the 'area effects' hypothesis, ABIs have been subjected to a detailed analysis. Although there are different emphases, one argument criticises the inward-looking nature of ABIs – that is, within the defined area – rather than their joined-up connection with other levels of society and the state. As with many other initiatives beyond those that are area-specific (e.g. people-based), the current coalition government is cutting funding to such schemes. Debate will subsequently centre on how this will shape the built environment and those that live within it.

Urban regeneration and renewal is another feature of urban policy that seeks to improve the attractiveness of place via social, economic and environmental means. For many cities, more recently the approach to urban regeneration has been property-led. For the UK, in outlining the historical development of property-led urban regeneration, between the 1940s and 1960s there was a transformation of the post-war legacy via slum clearance and a desire to end an acute shortage of housing. This transformation was in part aided in policy by comprehensive development areas (CDAs), a designation under the Town and Country Planning Act (1947) that allowed a local authority to acquire property in the designated area using powers of compulsory purchase in order to re-plan and develop urban areas suffering from war damage or urban blight (Johnson-Marshall, 2008). A return to redevelopment and rehabilitation returned in the early 1990s with policy approaches such as City Challenge, which have had some criticism (Fearnley, 2000) as has most (if not all) urban policies that have proceeded since (Cochrane, 2007). Urban regeneration strategies have more recently shifted to integrate a 'modernisation' strategy, which emphasise community responsibility rather than direct state provision. Modernisation has included a devolution of power to communities from the state by facilitating rather than providing (welfare) services, with a vision that communities and individuals can take responsibility for the conduct of their own lives (Raco and Imrie, 2003). For the UK, this 'modernisation' strategy can be more recently referred to as the 'Big Society' and 'Localism' policy agenda that in essence promotes a do-it-yourself ethos with respect to provision of public (or merit) goods and services where the market cannot provide for all citizens.

Interestingly, the UK coalition government has returned to some former low-tax incentivisation approaches that are attached to spatial boundaries (though not necessarily in urban areas) via the reintroduction of enterprise zones. This is in a similar vein to other spatially targeted economic development initiatives since the 1980s, that have had some element of policy transfer association with the US and its urban policy programmes (e.g. Empowerment Zones). In relation to spatially targeted economic development initiatives and environmental management, many British policy experiments have tried to export or import policy ideas to the United States. Examples of these transfers include: urban development grants (Goodhall, 1985); and urban development corporations (Raco, 2005); business improvement districts (Cook, 2008); and tax increment financing (Squires and Lord, 2012). In the case of enterprise zones, transfer has seen a UK to US direction of travel with individual US states enacting zone programmes in the early 1980s and the federal government adopting a zone programme in 1993 (Mossberger, 2000; Papke, 1993).

b. Economics in environmental policy

Environmental economic policy is concerned with meeting environmental problems at all spatial scales including those within urban spaces. At an international level (that encompasses all urban spaces), the recent United Nations Environment Programme (UNEP) green economy report (UNEP, 2011), the Organisation for Economic Co-operation and Development (OECD)'s interim green growth report (OECD, 2010) and the EU low carbon roadmap (European Commission, 2011) are the prime examples of environmental economic thinking, which is possible to trace back to the 1970s. From the 1970s, European governments began to get tough on pollution, whilst the majority of developing countries had little interest – viewed in part as a form of protectionism for the north. New environmental regulations were used as excuses for trade restrictions on the import of their products to northern markets. Foreign assistance increasingly directed itself away from the business of development toward environmental protection; and many developing countries felt strongly that the environment was a problem for the rich (Runnals, 2011).

By the 1980s virtually all OECD countries had established environment ministries and had passed legislation in the field. Despite this, it became apparent that little had changed in the developing world and the situation in some countries was becoming catastrophic (Runnals, 2011). This led to the creation of the World Commission on Environment and Development (WCED, 1987). The Brundtland Report's (1987) main insight was one that the world's critical ecosystems were under severe stress and some had passed vital thresholds and might never recover. Climate change and biodiversity loss were potentially catastrophic and immediate action was needed.

The Brundtland Report led eventually to the Earth Summit in Rio in 1992. At Rio a bargain was struck between north and south, with the rich countries gaining some acceptance of their agenda of climate change and biodiversity loss, deforestation and the destruction of migratory fish stocks. In exchange, developing countries received assurances of concessions on their agenda of increased development aid and better access to northern markets for their goods and debt relief (Runnals, 2011). After Rio, negotiations proceeded slowly until the development of the Kyoto Protocol in 1997. The Kyoto Protocol imposed real obligations on countries with mixed results. The United States opted out, and Europe began an important emissions trading scheme that placed a value on carbon and made it into a tradable commodity. The UNEP has just released its Green Economy Report (UNEP, 2011) plus the OECD has published its own Interim Green Growth Strategy (OECD, 2010).

There are plenty of signs that a new, lower-carbon economy is emerging without any help from the international political community (Runnals, 2011). Leadership in environmental resource allocation (particularly energy) is shifting to the BRIC (Brazil, Russia, India and China) countries in the same way that mainstream economic power is shifting. China is already host to the largest solar energy company in the world and is aggressively pursuing ambitious renewable energy targets.

Global environmental policy that will affect urban areas from an economic perspective are those such as the push for green city status projects and eco-cities. For instance, Freiburg im Breisgau (Germany) is often referred to as green city and is one of the few cities with a green mayor and is known for its strong solar economy. In China, where many new cities are under construction, it will be important to develop with a much lower reliance on fossil fuels. Green cities involve a network of decentralised mixed-use settlements, or 'clusters' connected by high-speed public transport and broadband communications, and may be a more sustainable solution. Connected by areas of intensive agriculture and natural systems, this decentralised model allows for the recycling of nutrients and water, as well as greater use of local materials. Certain clusters may eventually evolve into centres that offer particular commercial and social facilities, such as high-tech healthcare and education. Information and communication technologies would minimise the need to travel, thus reducing carbon emissions (Head and Singleton, 2011).

Eco-cities are targeted as other developments that could contribute to reducing carbon emissions within urban areas. In the second half of the 2000s, eco-cities appear to have become something of a global, mainstream phenomenon, with countries and cities competing to take a lead in developing and applying new socio-technological innovations and thus bringing about the next generation of sustainable towns and cities. However, it is not immediately clear exactly how significant a phenomenon eco-cities have become, in terms of both global spread and policy significance. Furthermore, it remains to be seen what are the key defining characteristics of eco-cities, and indeed what distinguishes them from 'normal' cities (Joss, 2010). To provide contemporary examples, in 2009 President Sarkozy of France announced that Paris would become the 'first post-Kyoto eco-city', and the British government published its decision to build four new 'eco-towns' across England (CLG, 2008b). Note that in 2011 the UK coalition government announced that only one of the proposed eco-towns, in Oxfordshire, will now actually be built to the originally proposed standards. The other proposed eco-towns will only need to be built to meet current building requirements, applied to any new build dwelling. On the outskirts of Abu Dhabi is the construction of Masdar City, the self-proclaimed 'world's first carbon-neutral zero-waste' city, while China is reported to have embarked on an ambitious programme to build some 40 new eco-cities (Joss, 2010).

c. Holistic policy: integration of urban and environmental economics (sustainable 'urban' development)

Urban economics and environmental economics have often operated in separate policy silos. It is here that more holistic thinking on the economics of use and management in shared urban spaces can begin to improve both the environment and social well-being of its inhabitants and visitors. The three pillars (economic, social and environmental) of sustainable development policy have begun to go some way in taking action in a more joined-up and multidisciplinary way.

Sustainable development is a relatively new concept that has made headway in becoming the new mantra of thinking in academia and practice. Whilst appearing to be a noble aim in itself, sustainable development involves temporal variances of time that is either past, present, or future – as well as bringing in variants in periods of time that are either short-term or long-term. As such, sustainable development is therefore aware of how any development (e.g. human or property) should encompass both short-term benefits as well as implications for longer-term success.

As described earlier in this book, sustainability is a slippery concept, despite its good-natured aims. For instance, sustainability may support more conservative projects when a more progressive and radical direction is needed. We could have sustainable inequality and a sustainable unfair system of resource allocation. Sustainability in the context of this study of urban and environmental economics is with regards to certain phenomena such as poverty or loss of habitat continuing for a long period of time, or almost indefinitely.

The most widely known definition of sustainable development comes from the Brundtland Commission, which defined it as meeting the needs of the present without compromising the ability of future generations to meet their own needs (WCED, 1987). More specifically, sustainable development for shared urban space tends to concentrate on the term 'sustainable urban development'. The term was initially generated from discussion with the Urban 21 conference in Berlin (Urban 21, 2000) where the term was recognised as:

> Improving the quality of life in a city, including ecological, cultural, political, institutional, social and economic components without leaving a burden on the future generations. A burden which is the result of a reduced natural capital and an excessive local debt…the flow principle, that is based on an equilibrium of material and energy and also financial input/output, plays a crucial role in all future decisions upon the development of urban areas.
>
> (Urban Future 21, 2000)

The further use of sustainable urban development has moved forward particularly within EU Regional policy. This has been to enable improved joined-up thinking within and between cities and urban areas in the EU. The importance of sustainable urban development has been stated by the Commissioner for European regional policy, when he stated that:

> With over 70% of Europeans living in urban areas, cities and metropolitan areas are the motors of economic growth and home to most jobs. They play a key role as centres of innovation and the knowledge economy. At the same time, urban areas are the frontline in the battle for social cohesion and environmental sustainability. The development of disadvantaged urban areas is an important step in unleashing economic powers by creating more cohesive and attractive cities.
>
> (EC, 2009)

d. Connecting the urban and the environmental – sustainable urban development: appraisal, assessment and evaluation

In practice, one way to unify both urban and environmental economics as an applied discipline is to use some of the approaches taken by 'sustainable urban development', particularly because sustainable urban development will consider environmental, social and economic components that tie in with the three main strands of this book – namely: (1) the build urban environment; (2) urban issues; and (3) environmental resource distribution.

With respect to sustainable urban development, the various concepts and theories need to be drawn together as method to make any meaningful change in policy and practice. This drawing together of all considerations in urban and environmental economics can be made via processes such as appraisal and assessment (as a form of hypothetical thinking into the future) and evaluation (as a form of retrospective, past, reflective thinking). As illustrators, this text has already referred to an appraisal technique such as cost-benefit analysis (CBA) and assessment in particular projects via methods such as environmental impact assessment (EIA). Evaluation can be thought of in examples such as the evaluation of large-scale urban development programmes. The key issues surrounding such appraisals and assessments with respect to urban space and resources (human and natural) via the sustainable development approach need to be introduced further, particularly in order to consider the merits and weaknesses of appraisals and assessments when used in more detail for policy and practice.

Appraisal – features

Economic appraisal and economic evaluation are general names for a set of techniques that weigh up the costs of an action against the benefits it provides. The distinction between appraisal and evaluation is that appraisal is undertaken before the action is taken, to decide what is to be done, and evaluation is undertaken after the action, to monitor its effects. More specifically, economic appraisal is a key tool for achieving value for money and is a systematic process for examining alternative uses of resources, focusing on assessment of needs, objectives, options, costs, benefits, risks, funding, affordability and other factors relevant to decisions. Appraisal is:

- designed to assist in defining problems and finding solutions that offer the best value for money (VFM);
- a way of thinking expenditure proposals through, right from the emergence of the need for a policy, programme or project, until its implementation;
- the established vehicle for planning and approving public expenditure policies, programmes and projects.

Good appraisal leads to better decisions and VFM. It facilitates good project management and project evaluation. Appraisal should not be optional; it is an essential part of good financial management, and it is vital to decision-making and accountability. Its principles must be applied, with proportionate effort, to all spending decisions, including small expenditures. The three main types of appraisal are CBA, cost-effective analysis and a scoring and weighting analysis.

Assessments – features

With regards to assessments, they are of most relevance here in terms of economic impact assessments (EcIA), environmental impact assessments (EnIA), or, more broadly, as previously discussed for urban and environmental economics, impact assessments (IA). Economic impact assessments (EcIAs) are concerned with identifying, measuring and quantifying the change in output associated with a public-sector intervention. Economic impact assessments (EcIAs) differ from cost-benefit analysis (CBA) in that the focus of the former is on changes to factors such as employment and output; e.g. gross value added (GVA), where the latter is more concerned with changes in social welfare. In addition, CBA is concerned with the opportunity cost of scarce resources and includes consideration of impacts with a market and non-market values. Note that what is being assessed with regards to the timing of an impact could be for activity that is past, current or into the future (but predominantly, assessment of the present) – it is more concerned with measuring or valuing the nature, quality or ability of someone or something. This means that assessments are often used interchangeably with appraisals (looking forward to see what is to be done in the future) and evaluation (looking more retrospectively at how an activity has fared, and hence whether the activity should continue).

Environmental impact assessment (EnIA) is a process that identifies the environmental effects (both negative and positive) of development proposals. It aims to prevent, reduce and offset any adverse impacts. Within the overall EnIA process there are two main stages. In the first stage the applicant undertakes an assessment so that environmental issues can be taken into account during the design of the project. This involves consultations, data collection and environmental studies to identify the effects and propose mitigation measures to prevent, reduce and offset them. This is reported in an environmental statement (ES), which is submitted in conjunction with the planning application. The planning authority then undertakes the second main stage by critically evaluating the statement, seeking further information from the applicant if necessary and taking into account additional consultations and public representations. This is to ensure the planning authority has sufficient reliable information to understand the likely environmental effects and specify any mitigation measures before the planning application is determined.

The types of project for which an EnIA has been undertaken are typically complex, with a wide range of environmental effects. They often occupy extensive sites and are in sensitive locations. They are likely to raise issues that are not always easy to resolve and that often attract contentious representations. For such projects, EnIA provides a systematic approach to obtaining and considering environmental information. For the overwhelming majority of development projects however, normal planning powers are perfectly adequate to gain environmental information and EnIA is not required. For an impact assessment (IA) that will consider all environmental, economic and social attributes, it is more generally about judging the effect that a policy or activity will have on people or places, and thus making a prediction or estimation of the consequences of a current or proposed action.

Evaluation – features

To evaluate is to judge or calculate the quality, importance, amount or value of something. This judgement or calculation is often more accurate given improved appropriation of current or past data, information and knowledge. A more formalised evaluation of an activity or policy can be described as a systematic, rigorous and meticulous application of scientific methods to assess the design, implementation, improvement or outcomes of a programme. It is a resource-intensive process, frequently requiring resources, such as, evaluator expertise, labour, time and a sizeable budget (Rossi *et al.*, 2004).

Economic evaluation, therefore, centres more on the methods to determine the value of a good, service, activity, policy, programme or project, albeit with a more retrospective approach. Even though an evaluation is more focused on the past account of activity and policy (e.g. programmes) that may have generated change, the techniques used are similar to those used in appraisals and assessments. For instance, techniques include the following:

- Cost-effectiveness compares the costs of different options for achieving a specific objective, such as building a particular road or meeting a greenhouse gas-emission reduction target. The quantity of outputs (benefits) are held constant, so there is only one variable, the cost of inputs.
- Cost-benefit analysis compares total incremental benefits with total incremental costs. It is not limited to a single objective or benefit. For example, alternatives may differ in construction costs and the quality of service they provide.
- Life cycle cost analysis is cost-benefit analysis that incorporates the time value of money. This allows comparisons between alternatives that provide benefits and costs at different times. For example, one option may cost more but be quicker to implement than another.
- Least cost planning is a type of benefit-cost analysis that considers demand management on equal terms with capacity expansion.
- Multiple accounts evaluation incorporates both quantitative and qualitative criteria and can be used when some impacts cannot be monetised or to allow decision-makers to evaluate each impact.
- Physical impacts and outcome techniques measure themes such as health, longevity, education levels, crime and personal satisfaction with life, without converting them into monetary values.

Economic evaluation in the use and management of shared urban space can be adopted. It can help guide decisions toward optimality, which refers to maximum social benefit. Economic evaluation involves quantifying incremental (also called marginal) economic impacts (benefits and costs) to determine net benefits or net value (benefits minus costs), and the distribution (also called incidence) of these impacts. Economic evaluation is not limited to market (measured in monetary units) impacts, it can also incorporate non-market resources such as personal time, health and environmental quality. Any good that somebody values is an economic resource, including non-market goods. Where units can be measured and quantified, the use of data and information can be used and presented to good effect in appraisals, assessments or evaluations. This use of data is where economic informatics can be extremely powerful in urban and environmental economics, and in doing so enable better decision-making.

e. Informatics and decision-making

Informatics studies the structure, algorithms, behaviour and interactions of natural and artificial systems that store, process, access and communicate information. It also develops its own conceptual and theoretical foundations and utilises foundations developed in other fields such as economics. Since the advent of computers, individuals and organisations increasingly process information digitally. This has led to the study of informatics that has computational, cognitive and social aspects, including study of the social impact of information technologies. For economics within the social sciences, the use of ICT and informatics has enabled analysis of urban areas and environmental resources to more sophisticatedly analyse and model relationships in space for particular selected variables, factors and components. For instance, urban change can be more intricately measured and modelled to understand (or maybe even project into the future), say, the impact of tragic congestion pollution on health over time.

As well as more sophisticatedly demonstrating what is occurring with respect to resource allocation in shared urban spaces, informatics can aid in providing better decision-making. Decision-making, in policy and practice, has greater chances of success with the use of good-quality data and information. For the study of economics, the development of ICT over the last decade has certainly improved the modelling of how natural and human resources are allocated over space. Statistical tools such as SPSS have enabled complex modelling of many variables to show relationships, such as through correlations and regression modelling. Furthermore, spatial tools such as geographic information systems (GIS), for instance, can now powerfully track economic changes at specific urban and rural locations. For instance, they can spatially layer data and information such as residential property prices, either statically as a snapshot or more dynamically over time.

With regards to construction and building in urban development, new integrated procurement methods are beginning to change the economic landscape via information technology communications (ICT). Building information management (BIM), for instance, is a relatively new approach to building design, construction and operation and is intended to change the way industry professionals think about how technology can be applied to the built environment. The intention is that there should be immediate availability and access to project design scope, schedule and cost information that is accurate, reliable, integrated and fully co-ordinated – and hopefully move beyond single-package approaches such as computer aided design (CAD) or off-the-shelf software such as Argus or Prime as used by professionals such as architects, quantity surveyors, structural engineers, mechanical/electrical engineers, project managers and construction design managers. BIM may even improve integration with extensions of single-package approaches or in the use of virtual design and construction (VDC). VDC is, essentially, an extension of CAD in that the digital design is fed into project management software and business process software to assess the best method and sequence of construction and to assess realistic times and costs of construction. With respect to economics, VDC can assess economic impact in the development of the built environment in urban areas by developing quantitative models of both cost and value of capital investments, including the project as a whole, individual project elements and any incremental investments required to change the process.

In the future, BIM is expected to empower both design and construction professionals to work more collaboratively throughout the project delivery process, focusing their energy on more value-added functions such as client requirements, creativity and problem-solving, while computers do the tedious tasks of number crunching. Ultimately, this approach and technology has the potential to enable the seamless transfer of knowledge from asset planning through design, construction, facilities management and operation, into the various disposal options. Whilst all parties involved in design and construction stand to gain from the adoption of BIM, it is the clients who will potentially benefit the most through the use of the facilities model and its embedded knowledge throughout the economic life of the building. This potential can only be realised if the information contained in the model remains accessible and usable across the variety of technology platforms likely over a long period. Given the accelerating pace of technology, in 20 to 30 years our now state-of-the-art hardware and software applications will be outdated and obsolete. It is therefore essential that BIM is developed within a universal, open-data standard to allow full and free transfer of data among the various applications (Greenhalgh and Squires, 2011).

f. Contemporary challenges and events: looking to the future

More recent concerns for the study of urban and environmental economics centre on the winners and losers in the continued process of globalisation and developments in ICT, coupled with the two major issues of this generation in dealing with the causes and consequences of (a) rising poverty and inequality, and (b) environmental degradation (including global warming, climate change and loss of biodiversity).

Events: G20 + world environmental summits

Significant decisions that affect the economic system of allocation for both equality and the environment is at the annual G20 event. For instance, G20 leaders in Pittsburgh, 2009, recognised that 'inefficient fossil-fuel subsidies encourage wasteful consumption, distort markets, impede investment in clean energy sources and undermine efforts to deal with climate change', and committed to 'rationalise and phase out over the medium term inefficient fossil-fuel subsidies that encourage wasteful consumption' (IEA *et al.*, 2010). They also acknowledged the challenges ahead, notably the need to prevent adverse impacts on the poorest by providing targeted cash transfers and other poverty-alleviation mechanisms. One of the key topics on the agenda for G20 events of interest for poverty and climate change is the use of subsidies. At the Toronto summit (2010), 13 countries outlined implementation strategies for phasing out selected fossil-fuel subsidies. The remaining seven countries (Australia, Brazil, France, Japan, Saudi Arabia, South Africa and the United Kingdom) concluded that they have no inefficient fossil-fuel subsidies. Other important discussions within the G20 will hope to encourage reductions in the quarter of a billion dollars that the developed countries currently provide domestic agricultural producers at the expense of developing countries (Runnals, 2011).

More recently has been the United Nations Climate Change Conference (UNFCCC), Durban 2011, which brought together representatives of the world's governments, international organisations and civil society. The discussions sought to advance the implementation of the Convention and the Kyoto Protocol, as well as the Bali Action Plan, agreed at Conference of the Parties (COP) 13 in 2007, and the Cancun agreements, reached at COP 16 in December 2010 (UNFCCC, 2011). The conference in 2011 was officially referred to as the seventeenth session of the Conference of the Parties (COP 17) to the UNFCCC and the seventh session of the Conference of the Parties serving as the meeting of the parties (CMP 7) to the Kyoto Protocol. The broad outcomes were in the creation of a Green Climate Fund for which a management framework was adopted. The fund is to distribute US$100 billion per year to help poor countries adapt to climate impacts. Further work to reduce global warming by two degrees is stressed and continues to be of issue in conferences and summits such as Rio+20 in 2012.

The United Nations Conference on Sustainable Development, also known as Rio 2012 or Rio+20, hosted by Brazil in Rio de Janeiro, was a 20-year follow-up to the historic 1992 United Nations Conference on Environment and Development (UNCED) that was held in the same city. Relevance to urban and environmental economics could not be any more pertinent with the conference having two central themes agreed upon by the member states that involve: (1) the green economy within the context of sustainable development and poverty eradication; and (2) institutional framework for sustainable development.

Summary

1   Linked to 'area effects' ideas, area-based initiatives (ABIs) have been critiqued as inward-looking – that is, within the defined area, rather than their joined-up connection with other levels of society and the state.
2   Urban regeneration and renewal is another feature of urban policy where between the 1940s and 1960s there was a transformation of the post-war property and infrastructure legacy via slum clearance.
3   There has been a return to redevelopment and rehabilitation in the early 1990s that shifted to integrate a 'modernisation' strategy.

4   There is a more recent return to some former low-tax incentivisation approaches that are attached to spatial boundaries (though not necessarily in urban areas).

5   Regarding targeted economic development initiatives and environmental management, examples of these include: enterprise zones; urban development grants; urban development corporations; business improvement districts; and tax increment financing.

6   For environmental policy, from the 1970s European governments were beginning to get tough on pollution, whilst the majority of developing countries had little interest – viewed in part as a form of protectionism for the north.

7   In the 1980s it became apparent that little had changed in the developing world (natural) environment and the situation in some countries was becoming catastrophic. This led to the creation of the World Commission on Environment and Development (the Brundtland Commission and the Brundtland Report).

8   In the 1990s a bargain was struck between north and south, with the rich countries gaining some acceptance of their agenda of climate change and biodiversity loss, deforestation and the destruction of migratory fish stocks. In exchange, developing countries received assurances of concessions on their agenda of increased development aid, better access to northern markets for their goods, and debt relief. Rio shifted the climate change negotiations leading to the Kyoto Protocol.

9   In the 2000s, Europe began an important emissions trading scheme that placed a value on carbon and made it into a tradable commodity. There was a significant Green Economy Report and Green Growth Strategy.

10   In the 2010s, there are many signs that a new, lower-carbon economy is emerging without any help from the international political community. It is interesting to see how leadership in this area is shifting to the BRIC countries in the same way that mainstream economic power is shifting. Especially significant or urban areas are the green city status and eco-city projects.

11   The further use of sustainable urban development has moved forward particularly within EU regional policy. This has been to enable improved joined-up thinking within and between cities and urban areas in the EU.

12   Economic appraisal is a key tool for achieving value for money and is a systematic process for examining alternative uses of resources, focusing on assessment of needs, objectives, options, costs, benefits, risks, funding, affordability and other factors relevant to decisions. The three main types of appraisal are CBA, cost-effective analysis, and a scoring and weighting analysis.

13   With regards to assessments, economic impact assessments (EcIAs) are concerned with identifying, measuring and quantifying the change in output associated with a public-sector intervention. Environmental impact assessment (EnIA) is a process that identifies the environmental effects (both negative and positive) of development proposals. It aims to prevent, reduce and offset any adverse impacts. This is reported in an environmental statement (ES), which is submitted in conjunction with the planning application.

14 Economic Evaluation therefore centres more on the methods to determine the value of a good, service, activity, policy, programme or project, albeit with a more retrospective approach. The techniques used are similar to those used in appraisals and assessments. These include: (1) cost-effectiveness; (2) cost-benefit analysis; (3) lifecycle cost analysis; (4) least cost planning; (5) multiple accounts evaluation; (6) physical impacts and outcomes.

15 The use of ICT and informatics have enabled analysis of urban areas and environmental resources to more sophisticatedly analyse and model relationships in space for particular selected variables, factors and components.

16 As examples of information communication technologies (ICT) development: (1) Statistical tools such as SPSS have enabled complex modelling of many variables to show relationships such as through correlations and regression modelling; (2) geographic information science (GISc), for instance, can now powerfully track economic changes at specific urban and rural locations; (3) building information management (BIM), for instance, is a relatively new approach to building design, construction and operation and is intended to change the way industry professionals think about how technology can be applied to the built environment.

17 Decision-making, in policy and practice, has greater chances of success with the use of good-quality data and information.

18 Recent concerns for the study of urban and environmental economics centre on the winners and losers in the continued process of globalisation and developments in ICT, coupled with the two major issues of this generation in dealing with the causes and consequences of (a) rising poverty and inequality, and (b) environmental degradation (including global warming, climate change and loss of biodiversity).

19 Emerging from UNFCCC 2011 were the creation of a Green Climate Fund for which a management framework was adopted. The fund is to distribute US$100 billion per year to help poor countries adapt to climate impacts. The key theme of Rio 2012 was the green economy within the context of sustainable development and poverty eradication.

References and further reading

Balchin, P., Isaac, D. and Chen, J. (2000) *Urban Economics: A Global Perspective*. New York: Palgrave.

Begg, D., Fischer, S. and Dornbusch, R. (1994) *Economics*. McGraw-Hill.

Blacksmith Institute (2007) *The World's Most Polluted Places*.

Bramley, G., Munro, M. and Pawson, H. (2004) *Key Issues in Housing: Policies and Markets in 21st C. Britain* Basingstoke: Palgrove MacMillan.

Brown, D. (1974) *Introduction to Urban Economics*. New York: Academic Press Inc.

Burgess, E.W. (1923) 'The growth of the city', in R.E. Parks and E.W. Burgess (eds) *The City*. Chicago: University of Chicago Press.

Button, K. (1976) *Urban Economics*. London: MacMillan Press.

C40 Cities (2011) C40 Cities – Climate Change Group homepage. http://www.c40cities.org/ [accessed 5 July 2011].

Centre for Cities (2011) *Cities Outlook 2011*.

CLG (Department of Communities and Local Government) (2006) Code for Sustainable Homes. CLG.

CLG (Department of Communities and Local Government) (2007) Sustainable Communities Act 2007: A Guide.

CLG (Department of Communities and Local Government) (2008a) *Definition of Zero Carbon Homes and Non-domestic Buildings*.

CLG (Department of Communities and Local Government) (2008b) *Eco-towns: Living a Greener Future*.

CLG (Department of Communities and Local Government) (2011) Localism Bill.

Club of Rome (1968) homepage. http://www.clubofrome.org/

Club of Rome (2009) *A New Path for World Development Programme: 2009–2012*.

Coase, R. (1960) 'The problem of social cost'. *Journal of Law and Economics*, 3, 1–44.

Cochrane, A. (2007) *Understanding Urban Policy: A Critical Approach*. Blackwell Publishing.

Cole, M. (1999) 'Limits to growth, sustainable development and environmental Kuznets curves: an examination of the environmental impact of economic development'. *Sustainable Development*, 7, 87–97.

Cook, I. (2005) 'Mobilising urban policies: the policy transfer of US business improvement districts to England and Wales'. *Urban Studies*, 45, 4, 773–95.

Crowley, T.J. (2000) 'Causes of climate change over the past 1000 years'. *Science*, 289, 279–90.

Daly, H. (1986) Comments on 'population growth and economic development.' *Population and Development Review*, 12, 583–5.

DECC (Department of Energy and Climate Change) (2011) *Carbon Capture and Storage*.

DEFRA (Department for Environment Food and Rural Affairs) (2011) *Sustainable Development*.

Denison, E. (1985) *Trends in American Economic Growth, 1929–1982*. Washington, DC: Brookings Institution.

DoE (Department of the Environment) (1975) National Land Use Classification: A Report of the Joint Local Authority, Local Authorities Management Services and Computing Committee, Scottish Development Department and DoE Study Team. HMSO.

EC (European Commission) (2009) *Promoting Sustainable Urban Development in Europe: Achievements and Opportunities*.

EC (European Commission) (2011) *Sustainable Use of Natural Resources*. European Commission: Environment.

Evans, A. (1985) *Urban Economics: An Introduction*. Oxford: Blackwell.

Fearnley, R. (2000) 'Regenerating the inner city: lessons from the UK's City Challenge experience'. *Social Policy and Administration*, 34, 5, 567–83.

Financial Times (2011) *The Financial Times Lexicon*.

Finkenrath, M. (2011) *Cost and Performance of Carbon Dioxide Capture from Power Generation*. International Energy Agency.

Garnett, D. and Perry, J. (2005) *Housing Finance*. Coventry: CIH.

Gilpin, A. (2000) *Environmental Economics: A Critical Review*. Chichester: Wiley.

Goodall, G. (1972) *The Economics of Urban Areas*. Oxford: Pergamon Press.

Goodhall, P. (1985) 'Urban development grant – an early assessment of regional implications'. *Journal of Environmental Planning and Management*, 28, 1, 40–2.

Greenhalgh, B. and Squires, G. (2011) *Introduction to Building Procurement*. Abingdon: Spon Press.

Haig, M. (1926) 'Toward an understanding of the metropolis', *Journal of Economics*, 40, 179–208.

Harris, C. and Ullman, E. (1945) 'The nature of cities'. *The Annals of the American Academy of Political and Social Science*, 242, 7–17.

Head, P. and Singleton, D. (2011) 'Green cities of the future', *The Guardian*. http://www.guardian.co.uk/sustainable-business/cities-self-sufficient-new-urban-energy-centres [accessed February 2012].

Hoyt, H. (1939) *The Structure and Growth of Residential Neighbourhoods in American Cities Washington*. Federal Housing Administration.

Huf Haus. (2012) The Huf Haus Company. http://www.huf-haus.com/

ICLEI – Local Governments for Sustainability (2012) homepage. http://www.iclei.org/ [accessed Feb 2012].

IFCDI (2011) Xinhua-Dow Jones IFCDI International Financial Centers Development Index.

IIED (2007) 'Adapting to climate change in urban areas: the possibilities and constraints in low- and middle-income nations', in Human Settlements Discussion Paper Series. Theme: Climate Change and Cities – 1. International Institute for Environment and Development.

Joardar, S. (1998) 'Carrying capacities and standards as bases towards urban infrastructure planning in India: a case of urban water supply and sanitation'. *Habitat International*, 22, 3, 327–37.

Johnson-Marshall, P. (2008) *Rebuilding the City: The Percy Johnson-Marshall Collection*. University of Edinburgh.

Joss, S. (2010) 'Eco-cities – a global survey 2009'. *WIT Transactions on Ecology and the Environment*, 129, 239–50.

JRF (Joseph Rowntree Foundation) (2009) A Minimum Income Standard for Britain in 2009.

King, P. (2008) *Understanding Housing Finance*. Abingdon: Routledge.

Kirchner, J., Leduc, G., Goodland, R. and Drake, J. (1985) 'Carrying capacity, population growth, and sustainable development', in D. Mahar (ed.) *Rapid Population Growth and Human Carrying Capacity: Two Perspectives*. Staff Working Papers 690, Population and Development Series. Washington, DC: The World Bank.

Kohler, H. (1973) *Economics and Urban Problems*. Lexington, MA: D.C. Heath.

Kovarik, W. (2011) Environmental History Timeline [accessed May 2011].

LeGrand, J., Propper, C. and Smith, S. (2008) *The Economics of Social Problems*. Basingstoke: Palgrave Macmillan.

Lipsey, R. and Chrystal, A. (1995) *An Introduction to Positive Economics*. Oxford: Oxford University Press.

Malthus, T.R. (1798) *Essay on the Principle of Population as it Affects the Future Improvement of Society*. London: Ward Lock.

Mankiw, N. Gregory (2007) *Macroeconomics*. New York: Worth.

Marshall, A. (1920) *Principles of Economics*, 8th edn. London: Macmillan and Co.

McDonald, J. and McMillen, D. (2008) *Urban Economics and Real Estate*. Oxford: Blackwell.

McKenzie, R., Park, R. and Burgess, E. (1967) *The City*. Chicago: University of Chicago Press.

Meadows, D., Randers, J. and Behrens, W. (1972) *The Limits to Growth*. New York: Universe Books.

Mill, J.S. (1848) *Principles of Political Economy*. New York: Appleton.

Mossberger, K. (2000) *The Politics of Ideas and the Spread of Enterprise Zones*. Washington, DC: Georgetown University Press.

Natural England (2008) *State of the Natural Environment 2008*. A Report.

NEAA (Netherlands Environmental Assessment Agency) (2008) *Global CO_2 Emissions: Increase Continued in 2007*. PBL NEAA.

NHBC Foundation. (2009) *Zero Carbon Homes – An Introductory Guide for Housebuilders*.

OECD (2010) *Interim Report of the Green Growth Strategy: Implementing Our Commitment for a Sustainable Future*. Organisation for Economic Co-operation and Development.

ONS (Office for National Statistics) (2001) Census 2001: Definition of Urban Areas. HMSO. http://www.statistics.gov.uk/census2001/pdfs/urban_area_defn.pdf [accessed February 2012].

Papke, L. (1993) 'What do we know about enterprise zones'. *Tax Policy and the Economy*, 7, 37–72.

Park, R.E. and Burgess, E.W. (1925) *The City: Suggestions for Investigation of Human Nature in the Urban Environment*. Chicago, IL: University of Chicago Press.

Pearce, D., Kerry, T. and Bateman, I. (2000) *Environmental Economics: An Elementary Introduction*. Prentice-Hall.

Pearce, D., Markandya, A. and Barbier, E.B. (1990) *Blueprint for a Green Economy*. London: Earthscan Publications.

Planning Portal (2011) Shapps confirms revised definition for zero-carbon homes. Planning Portal.   http://www.planningportal.gov.uk/general/news/stories/2011/may11/19may11/190511_5 [accessed February 2012].

POA (Parliament of Australia) (2008) Climate change and global warming – What's the difference? Parliament of Australia Library. http://www.aph.gov.au/library/pubs/climatechange/theBasic/climate.htm [accessed 5 July 2011].

Raco, M. (2005) 'A step change or a step back? The Thames Gateway and the rebirth of the urban development corporations'. *Local Economy*, 2, 141–53.

Raco, M. and Imrie, R. (2003) *Urban Renaissance? New Labour, Community and Urban Policy*. Bristol: Policy Press.

Rees, W. (1996) 'Revisiting carrying capacity: area-based indicators of sustainability in population and environment'. *Journal of Interdisciplinary Studies*, 17, 3. Human Sciences Press, Inc.

Ricardo, D. (1817) *Principles of Political Economy and Taxation*. London: Pelican Books.

Rosenthal, S. and Strange, W.C. (2003) 'Geography, industrial organization, and agglomeration'. *Review of Economics and Statistics*, 85, 2, 377–93.

Rossi, P.H., Lipsey, M.W. and Freeman, H.E. (2004) *Evaluation: A Systematic Approach*, 7th edn. Thousand Oaks, CA: Sage.

Runnals, D. (2011) 'Environment and economy: joined at the hip or just strange bedfellows?' S.A.P.I.EN.S., 4, 1. http://sapiens.revues.org/1150

Santer, B.D., Taylor, K.E., Wigley, T.M.L. *et al.* (1996) 'A search for human influences on the thermal structure in the atmosphere. *Nature*, 382, 39–46.

Sawyer, M. (ed.) (2005) *The UK Economy*. Oxford: OUP.

Smith, A. (1776) *An Enquiry into the Nature and Cause of the Wealth of Nations.*

Squires, G. and Lord, A. (2012) 'The transfer of tax increment financing (TIF) as an urban policy for spatially targeted economic development initiatives'. *Land Use Policy*, 29, 4, 817–26.

Stern, N. (2006) *Stern Review on the Economics of Climate Change* (pre-publication edition). Executive Summary. London: HM Treasury.

Stivers, R. (1976) *The Sustainable Society: Ethics and Economic Growth*. Philadelphia, PA: Westminster Press.

Sweeney, J. (2000) 'Economics of energy'. *Encyclopedia of the Social and Behavioral Sciences*, 4.9, article 48.

Turner, R., Pearce, D. and Bateman, I. (1994) *Environmental Economics: An Elementary Introduction*. New York: Harvester Wheatsheaf.

UN Habitat (2011) *Cities and Climate Change – Global Report on Human Settlements.*

UN (2005) *World Urbanization Prospects* (2005 revision). Department of Economic and Social Affairs, United Nations.

UNEP (United Nations Environment Programme) (2002) *State of the Environment and Policy Retrospective: 1972–2002.*

UNEP (United Nations Environment Programme) (2011) *Green Economy Report.*

UNFCCC (2011) United Nations Framework Convention on Climate Change homepage http://unfccc.int/2860.php [accessed 2012].

US Future 21 (2000) *Urban Future 21: A Global Agenda for 21st Century Cities*. Abingdon: Spon Press.

US Census Bureau (2007) US Census Bureau: State Income Data. http://www.census.gov

Van den Bergh, J. C. J. M. (2000) 'Ecological economics: themes, approaches, and differences with environmental economics'. Tinbergen Institute Discussion Paper.

WCED (1987) *Our Future: Report of the World Commission on Environment and Development*. Brundtland Commission, formerly the World Commission on Environment and Development (WCED).

Winger, A. (1977) *Urban Economics: An Introduction*. Colombus, OH: Merrill.

World Bank (2000) *Entering the 21st Century: World Development Report 1999/2000*. New York, Oxford University Press.

World Mayors Council (2011) *World Mayors Council on Climate Change.* http://www.worldmayorscouncil.org/ [accessed 5th July 2011].

Worster, D. (1977) *Nature's Economy: A History of Ecological Ideas*. Cambridge: Cambridge University Press.

WRI (World Resources Institute) (1996) *World Re-sources 1996–1997: A Guide to the Global Environment*. World Re-sources Institute, UN Environment Programme, UNDP and the World Bank, Washington, DC.

Index

Milton Keynes UK
Ingram Content Group UK Ltd.
UKHW040100071024
449327UK00019B/701